AF782786

Managing Famine, Flood and Earthquake in China

China suffers frequently from many types of natural disasters, which have affected the lives of many millions of Chinese. The steps which the Chinese state has taken to prevent disasters, mitigate their consequences, and reconstruct in the aftermath of disasters are therefore key issues. This book examines the single metropolis of Tianjin in northern China, a city which has suffered particularly badly from natural disasters – the great famine of 1958–61, the great flood of 1963 and the great earthquake of 1976. It discusses how the city managed these disasters, what policies and measures were taken to prevent and mitigate disasters, and to promote reconstruction afterwards. It also explores who suffered from and who benefited from the disasters. Overall, the book shows how disaster management was erratic, sometimes managed highly efficiently and in other cases disappointingly delayed and inept. It concludes that, although the Maoist state possessed formidable resources, disaster management was always constrained by other political and economic considerations, and was never an automatic priority.

Lauri Paltemaa is Professor of East Asian Politics and Contemporary History and Director of the Centre for East Asian Studies, University of Turku, Finland.

Routledge Studies in the Modern History of Asia

For a complete list of titles in this series, please visit www.routledge.com

84. China and the First Vietnam War, 1947–54
Laura M. Calkins

85. The Jesuit Missions to China and Peru, 1570–1610
Ana Carolina Hosne

86. Macao – Cultural Interaction and Literary Representations
Edited by Katrine K. Wong and C. X. George Wei

87. Macao – The Formation of a Global City
Edited by C. X. George Wei

88. Women in Modern Burma
Tharaphi Than

89. Museums in China
Materialized power and objectified identities
Tracey L-D Lu

90. Transcultural Encounters between Germany and India
Kindred spirits in the 19th and 20th centuries
Edited by Joanne Miyang Cho, Eric Kurlander and Douglas T. McGetchin

91. The Philosophy of Japanese Wartime Resistance
A reading, with commentary, of the complete texts of the Kyoto School discussions of "The Standpoint of World History and Japan"
David Williams

92. A History of Alcohol and Drugs in Modern South Asia
Intoxicating affairs
Edited by Harald Fischer-Tiné and Jana Tschurenev

93. Military Force and Elite Power in the Formation of Modern China
Edward A. McCord

94. Japan's Household Registration System and Citizenship
Koseki, identification and documentation
Edited by David Chapman and Karl Jakob Krogness

95. Itō Hirobumi – Japan's First Prime Minister and Father of the Meiji Constitution
Kazuhiro Takii

96. The Non-Aligned Movement and the Cold War
Delhi – Bandung – Belgrade
Edited by Nataša Mišković, Harald Fischer-Tiné, and Nada Boškovska

97. The Transformation of the International Order of Asia
Decolonization, the Cold War, and the Colombo Plan
Edited by Shigeru Akita, Shoichi Watanabe and Gerold Krozewski

98. Xinjiang and the Expansion of Chinese Communist Power
Kashghar in the early twentieth century
Michael Dillon

99. Colonial Counterinsurgency and Mass Violence
The Dutch Empire in Indonesia
Edited by Bart Luttikhuis and A. Dirk Moses

100. Public Health and National Reconstruction in Post-War Asia
International influences, local transformations
Edited by Liping Bu and Ka-che Yip

101. The Pacific War
Aftermaths, remembrance and culture
Edited by Christina Twomey and Ernest Koh

102. Malaysia's Defeat of Armed Communism
The Second Emergency, 1968–1989
Ong Weichong

103. Cultural Encounters and Homoeroticism in Sri Lanka
Sex and serendipity
Robert Aldrich

104. Mobilizing Shanghai Youth
CCP internationalism, GMD nationalism and Japanese collaboration
Kristin Mulready-Stone

105. Voices from the Shifting Russo-Japanese Border
Karafuto / Sakhalin
Edited by Svetlana Paichadze and Philip A. Seaton

106. International Competition in China, 1899–1991
The Rise, Fall, and Eventual Success of the Open Door Policy
Bruce A. Elleman

107. The Post-war Roots of Japanese Political Malaise
Dagfinn Gatu

108. Britain and China, 1840–1970
Empire, Finance and War
Edited by Robert Bickers and Jonathan Howlett

109. Local History and War Memories in Hokkaido
Edited by Philip A. Seaton

110. Thailand in the Cold War
Matthew Phillips

111. Early Modern Southeast Asia, 1350–1800
Ooi Keat Gin and Hoang Anh Tuan

112. Managing Famine, Flood and Earthquake in China
Lauri Paltemaa

Managing Famine, Flood and Earthquake in China

Tianjin, 1958–85

Lauri Paltemaa

Routledge
Taylor & Francis Group
LONDON AND NEW YORK

First published 2016 by Routledge

2 Park Square, Milton Park, Abingdon, Oxfordshire OX14 4RN
711 Third Avenue, New York, NY 10017

Routledge is an imprint of the Taylor & Francis Group, an informa business

First issued in paperback 2017

Copyright © 2016 Lauri Paltemaa

The right of Lauri Paltemaa to be identified as author of this work has been asserted by her in accordance with sections 77 and 78 of the Copyright, Designs and Patents Act 1988.

All rights reserved. No part of this book may be reprinted or reproduced or utilised in any form or by any electronic, mechanical, or other means, now known or hereafter invented, including photocopying and recording, or in any information storage or retrieval system, without permission in writing from the publishers.

Notice:
Product or corporate names may be trademarks or registered trademarks, and are used only for identification and explanation without intent to infringe.

British Library Cataloguing in Publication Data
A catalogue record for this book is available from the British Library

Library of Congress Cataloging in Publication Data
Paltemaa, Lauri.
Managing famine, flood, and earthquake in China : Tianjin, 1958-1985 /
Lauri Paltemaa.
(Routledge studies in the modern history of Asia ; 112)
Includes bibliographical references and index.
1. Emergency management–China–Tianjin Region–History–20th century.
2. Crisis management–China–Tianjin Region–History–20th century.
3. China–Politics and government–1949- I. Title.
HV551.5.C6P35 2016
363.340951'15409045–dc23
2015016285

ISBN: 978-1-138-83139-1 (hbk)
ISBN: 978-1-138-47617-2 (pbk)

Typeset in Times New Roman
by Cenveo Publisher Services

Contents

Maps and tables

Maps

Tables

Acknowledgements

Writing this book began in 2006 as a part of the Modes of Governance in Urban China research project funded by the Academy of Finland. During this time, many people have helped in a number of ways, both academic and pragmatic. I would like to therefore express my gratitude to a number of them: Professor Hou Jie of the Nankai University History Department deserves special thanks, as does Doctor Ren Yunlan from the Tianjin Academy of Social Sciences together with my invaluable local research assistants Liu Wenhui, Zhao Tianlu and Doctor Qin Fang, who all have made this study possible in the first place. I would also like to thank the helpful staff of the Tianjin Academy of Social Sciences and the University Service Centre of the Chinese University of Hong Kong, whose archives I have been able to use in my work. I have received many useful comments and had helpful discussions with many scholars on Chinese disasters including Felix Wemhaeur, Jeremy Brown, Andrea Janku and Pierre Fuller, as well as numerous commentators in conferences and workshops I have attended under this topic. Timothy Glogan has helped me to proofread the final text, for which I owe him my gratitude. All mistakes in the text are of course entirely my own. The project has been funded not only by the Academy of Finland, but also the Emil Aaltonen Foundation and the Kone Foundation, for which I am also thankful. And, finally, I express my gratefulness to Professor Klaus Mühlhahn for setting the wheels turning in 2006. As the Chinese well know, perseverance is the key to victory.

Lauri Paltemaa
Turku, 7 April 2015

Abbreviations

CCP	The Chinese Communist Party
CEO	City Earthquake Office
CIFRC	China International Famine Relief
CR	The Cultural Revolution
GLF	The Great Leap Forward
HBFPHQ	Hebei Provincial Flood Protection Headquarters
KMT	Kuomintang
OEO	Organization Education Office
PLA	The People's Liberation Army
PRC	People's Republic of China
RERWHQ	Headquarters for Resisting Earthquake and Relief Work
ROC	Republic of China
SOE	State Owned Enterprise
TJFPHQ	Tianjin City Flood Prevention Headquarters
TR	Tianjin Ribao (Tianjin daily)

Introduction

This is a book about how Mao-era China coped with disasters. Many scholars who have researched natural disasters from historical, social-scientific or anthropological perspectives have noted how disasters function as great revealers of social and political relations in society.[1] Unfortunately for China, it has a lot to offer for research in this respect. According to the State Council's *White Paper* on disaster management (2009), more than half the Chinese population is situated in areas where serious earthquakes, meteorological, geological and/or marine disasters can occur. Floods can threaten two-thirds of China's land area. In the past, all the country's provinces have been struck by earthquakes of magnitude 5 or more on the Richter scale.[2] The human and economic costs of natural disasters have been staggering in the modern and contemporary era. Most destructive disasters have been earthquakes, floods and droughts. It has been estimated that between 1900 and 2008, earthquakes killed over 870,000 Chinese, affected the lives of some 68.5 million, and caused damage costing over 930 billion USD. During the same period, floods are estimated to have killed some 4.3 million people, affected the lives of 1.3 billion, and caused damage costing over 1.2 trillion USD.[3] In addition, droughts have contributed to repeated cycles of famines with myriad casualties up until the contemporary era in China.

Disasters concern governments for obvious reasons. A disaster can be seen as a chain of events, which affects large parts of a community in extremely negative ways. In disasters, things that support our lives, production and health fail on a large scale. They are about massive destruction, death, sickness, carnage and injury. They are extraordinary occasions where individual lives and properties are at stake; where whole communities may break down or be totally annihilated. Their scale, often their suddenness and the fact that they affect virtually everybody in a community make them an urgent matter for governance.[4]

The role of the state in disasters and disaster management is of course related to the way government's role in society is organized in general and what kind of expectations people therefore have for it. In China, as in the contemporary West, governments have been expected to protect and help their citizens in disasters that overwhelm their own abilities to cope with the situation. In China this prerogative may even have been stronger and older than in any other country in history.[5]

Apart from seeing disasters as extraordinary negative and massive collective events, disaster scholars tend to agree that disasters are social processes and should be studied as such. It is natural hazards that usually trigger disasters, but it is human activity before, during and after the hazard event that turns a change in natural environment into a social disaster. Sometimes the triggering cause can also be man-made, such as a policy decision or action from outside the affected community, such as military invasion.[6] Consequently, disaster management also needs to be understood as a combination of government and social processes, which aim to prevent hazards, mitigate their impact on communities and help communities to recover afterwards.[7]

Here it is useful to make a distinction between two commonly used concepts. 'Disasters' and 'crises' are sometimes used interchangeably in everyday talk and even in research, but in this book these words are reserved to mean different, although related, phenomena. A disaster here means an event of large-scale death, injury and/or economic dislocation. A crisis is a chain of events where the danger of a disaster is clear and present, but the processes that constitute a disaster have not yet started on a large scale. Death, injury and economic dislocation can also occur in crises, but not yet on a scale that disrupts the whole community, which is what a disaster does. A crisis may turn into a disaster, but if a crisis is properly managed (or by sheer luck), this outcome may be averted. If not, a disaster is a 'crisis with a bad ending', as noted by E.L. Quarantelli *et al.*[8]

This brings us to the main theme of the book. Throughout Chinese history, disasters have been managed in order to prevent them, mitigate their impacts and rebuild after them, sometimes skillfully, sometimes less so. This study is basically about how this work was done in Tianjin city in the Great Leap Forward famine, the great flood of 1963 and the great 'Tangshan' earthquake in 1976. By taking disaster management, or governance – both terms which are used interchangeably here to avoid too much repetition of the same phrase in the text – as the focus, the book approaches disasters as opportunities to conduct historical socio-political analysis of Maoist local government and society. By studying famine, flood and earthquake in Tianjin city, we reconstruct how Maoist urban government tried to prevent crises from turning into disasters, or failing this, to mitigate their impact on the population and to rehabilitate communities afterwards. Through this, we learn not only how the Maoist regime dealt with disasters, but also how it worked under crisis situations. Finally, we can ask if there is anything to learn from this experience for contemporary disaster management.

During the Mao-era, disaster governance had a very distinctive character. Indeed, one can arguably talk about a specific, Maoist type of campaign-based disaster governance. Its strengths were pronounced especially in damage mitigation during the hazards. This was essentially because the party-state was able to put its notable skills at mass mobilization into use in disaster management. But in many other aspects, such as taking preventive measures and in rebuilding after hazards, the system was less than perfect. Most crucially, although local metropolitan government was in principle highly capable of turning its machinery against natural disasters, the central government constituted the crucial bottleneck in the whole system, making disaster governance erratic and sometimes making it fail totally. The way disasters were managed had concrete consequences for the whole population of Tianjin.

We also look at how Maoist disaster management affected vulnerabilities. Individual death and injury in disasters is often a result of seemingly random, even freak circumstances. Someone barely avoids being crushed by a falling rooftop, while a man next to her ends up buried in the ruins during an earthquake. Someone lives higher on the ground and is not swept away with the rest of the village when the dykes of a nearby river suddenly collapse. After the flood, one person catches dysentery during an epidemic and dies, while a person living next door is unaffected. But when these deaths are analysed in a larger perspective, patterns of vulnerability usually emerge. Indeed, a basic finding in disaster research is that societies distribute vulnerability differently.[9] For example, sociological research on disasters shows that almost invariably richer people tend to be less vulnerable than the poor.[10] But in Maoist China wealth was not a factor in social divisions, except in a historical sense affecting people's class labels, so who were the vulnerable in Maoist society and why? As the pages to follow show, Maoist society distributed its vulnerabilities according to the largely artificial socio-political structures that the party-state created after 1949. However, on occasion effective disaster management could also greatly reduce these vulnerabilities, at least temporarily.

Maoist China is also a prime example of how a 'crisis rarely, if ever, "speaks for itself"'.[11] Leaders not only manage crises, they also impute them with meanings. The Maoist party-state was very skilled in putting a positive spin on disasters and had no moral compunction about lying to its people if needed. Notable energy was directed at governing the discursive dimension of disasters, and propaganda, cultural and emotional governance[12] were therefore an essential part of disaster governance in Maoist China. Different kind of disasters required different kinds of narratives, but the basic functions of disaster propaganda were always the same in Maoist China. They were conscious products of the Communist regime with the aim of making disasters lessons for the population, not for the regime, as they had been regarded in the past.

Fundamentally, studying Maoist disaster management shows how responding to disasters did not necessarily gain any priority in political decision-making. During the Mao-era the decisions to engage in active disaster management campaigns were always based on political and economic calculations. In this respect, Maoist disaster management followed Chinese tradition. As Pierre-Étienne Will and R. Bin Wong have shown, pre-Tang dynasty disaster management was already constrained by a 'tension between revenue-making and social welfare' over the issue of whether one should make profit when selling relief grain from granaries during lean years.[13] Kathryn Edgerton-Tarpley presents two examples of such calculations from the late Qing and Republican periods. During the Great North China famine in 1876–9, the Qing court considered the decision over famine relief a decision over whether to fund the army or relief aid. When, however, the Kuomintang (KMT) leaders made a decision to breach the dykes of the Yellow River to stop the advancing Japanese imperial army in 1938, it was a decision about defending the nation versus protecting individual people's lives.[14] In Maoist China the basic question was over allocating resources to disaster management or production, and as we will see in the coming pages, this basic dilemma weakened Maoist disaster management considerably.

However, in order to understand the features of Maoist disaster management, we also need to know something about the history of disaster governance in China, as well as the general socio-political structures and governance principles in Maoist society, both of which provide the basis whereon Maoist disaster management was built.

Brief Overview of Disasters and Disaster Management in Chinese History

Chinese historical records mention numerous types of disaster, but, arguably, famines and hazards related to famines play a prominent role in them. Famines are of course often follow-on disasters, meaning that they follow (which is not to be confused with causation) natural hazards such as drought, floods and locusts. Historical records of famines therefore also often mention these hazards. The first historical entry on floods can be found in *Da Yu zhishui* (Great Yu Controls the Waters, 2200–2100 BC), which eulogizes the flood control efforts by the legendary Sage Yu, the father of the founder of the legendary Xia dynasty, and shows how the creation of the early Chinese state formations was closely related to disaster management.[15]

Probably the first entry on famine by a historical dynasty is from Western Zhou mentioning severe drought and famines in 858–853 BC. There are different estimates on how many famines there have been in Chinese history. One account counts 381 mentions of a 'famine' in the dynastic records from the Western Han to the Qing dynasty, but smaller local famines were much more numerous than that.[16] So much so, that by the late nineteenth century China had earned a nickname of the 'Land of Famine' in the English-speaking world. Famines were a persistent and recurring state of affairs of agrarian society in imperial and republican eras, not extraordinary or exceptional events. But this is not to say that they were received with fatalist abandon, or that there was no attempt to manage them or the other kinds of disasters – quite the contrary. Indeed, as Lillian M. Li has argued 'no other civilization had such continuous tradition of thinking about famine, and no other nation's modern history has been so influenced by hunger and famine'.[17] This can be said to hold true with disasters in China in general.

Many reasons for persistent famines have been singled out. The monsoon climate creates cycles of droughts and floods in large parts of China exacerbated by what has been currently termed as the El Niño phenomenon, but socio-political factors have been equally important. These have included wars, bad government, population growth, underdeveloped communications, and low technological standards in agriculture. A direct human factor in causing disasters leading to famines has played its part from very early on, too. Historical records frequently mention ill-conceived or poorly maintained waterworks by officials and the gentry with misappropriation of flood control funds as causes of floods. Together with large parts of the population living just above the subsistence level, these factors created a large vulnerable peasant population, which lacked the resources for self-protection and often received too little social protection from dynastic governments for surviving recurrent crop failures. As late as the Republican period, it was estimated that 3–4 million people starved to death yearly and ten times that number were living on subsistence level where any natural hazard or social disturbance could plunge them under to starvation. Contemporary official Chinese history writing blames feudal exploitation for this.[18] However, the relation between the socio-political order and famines was less straightforward than this, since disaster management was also practiced at least as long as there had been historical records in China.

Throughout Chinese history, famines and famine prevention had an intimate connection to the rulers, who were regarded as being responsible for the well-being of their people. This was codified in the Mandate of Heaven (*tianming*) philosophy that was first developed during the Western Zhou dynasty. According to the early version of the Mandate of Heaven, God (*tiandi*, *shangdi*) ordered calamities as punishment for Man. The king, or later the emperor, was regarded as God's son and therefore having a direct connection to him. Later, God was replaced by more impersonal Heaven, but the thinking remained otherwise the same. The Western Zhou developed the notion that maintaining virtue (*de*) by the king was the main method of protecting people from the wrath of heaven. Protecting people (*bao min*) from disasters thus became the justification for the monarchy. It was not a coincidence that the first bureaucratic structure assigned to deal with famine relief was also recorded in the Western Zhou. Later the Confucian (*Ru*) school of political philosophy, and above all one of its leading philosophers, Mencius (372–289 BC), further elaborated the doctrine of the Mandate of Heaven in this respect. According to him, the ruler could lose this mandate if he neglected his duties of protecting the common people. Famines were at the heart of this as touchstones for a ruler's morality and efficacy.[19]

However, while the ruling Confucian elite agreed that disasters manifested the displeasure of Heaven, the answer for how to manage disasters was not simple. Was it enough for the rulers to cultivate their morals and thereby appease Heaven, or was more active disaster management needed? As Rothschild has shown, the Tang court became divided over this issue, when it had to decide on how to deal with a massive locust infestation that threatened parts of the realm with famine. The passive line saw moral cultivation as enough in itself, while activists in the court demanded affirmative action on locust extermination without denying the need for moral cultivation, as such.[20] Therefore, *Tianming* thinking did not translate itself automatically into disaster prevention and relief efforts, but the thought had to be interpreted thus first.

Mandate of Heaven thinking did not mean that the blame for disasters lay only with the emperor either. As Mark Elvin has shown, Qing emperors accepted the magical explanation that the negative energy from popular resentment could rise up to the heaven and cause disaster. On the other hand, if the emperor and/or local officials were able to approach heaven with a sincere spiritual attitude, they might be able to attain the heart-mind of heaven and appease its wrath. According to this high Qing thinking, not only could emperors and officials at large be held responsible for disasters, but also commoners. Disasters therefore became not only signs for the rulers to amend their erroneous ways of ruling, but also for common people to rectify their behaviour.[21] It is of course impossible to tell how widely such official discourse was accepted, but it is important to note how even the *Tianming* idea that originally entitled people to protection by the state could be arrogated by the elite and used to turn blame away from the emperor and his bureaucracy and on the people themselves.

The moral cultivation/action dilemma in disaster management was never solved on a theoretical level during the dynastic era. Instead, both methods were employed by rulers simultaneously. Different dynasties paid attention to disaster prevention and mitigation in different degrees, as well as conducting heaven-appeasing rituals when disasters struck. The Zhou dynasty had already developed an advanced bureaucratic rule by the standards of time, where disaster protection and relief work was emphasized and disaster management (*huangzheng*) became a function of bureaucracy. Measures to improve agricultural production were taken. For example, irrigation works, reservoirs, and levies were constructed to protect against drought and flooding and ensuing dynasties continued these practices. Attempts at early meteorology were also taken and a county-level granary system was established to appease popular sentiment in time of disaster and prevent people from taking refuge in other parts of the country. The *Book of Rites* recommended one-year storages per three crop years, and saw that having grain for three lean years was a minimum for the proper functioning of the state.[22]

By the time of the great emperors of the Qing dynasty in the eighteenth century, Chinese classical disaster governance had reached its peak and was arguably more advanced both practically and discursively than anywhere in the world, including the West at the time.[23] Apart from the emperor's moral cultivation, classical disaster relief practices included the government handing out active relief aid, regulating grain markets through 'ever constant' state granaries and utilizing local charity and community granaries.[24] Other methods included transferring grain from other areas to disaster-stricken ones, sheltering refugees and helping in their repatriation, and granting tax breaks or postponing taxation in disaster-stricken areas. Relief-for-work was a guiding principle, whereby disaster refugees could be made to build roads or river dykes or to weave baskets and straw mats in return for relief aid. Individuals would often take refuge in other parts of China as families or even as whole villages, or resort to more desperate measures such as selling their babies, eating ersatz food (or anything resembling it), banditry and even cannibalism. The tradition of such disaster prevention measures carried over to the republican era and could also be found in many disaster relief practices in the People's Republic.

The weakening of the late Qing state brought with it a number of changes to disaster management as well. The devastation wrought by large rebellions and wars in the mid-nineteenth century translated into weaker administration of civilian affairs, fewer resources for relief aid and the decay of many existing preventive measures such as river embankments. Recent research has established that by and during the Great North China Famine (1876–9) a new mixed type of disaster management model had emerged. The Qing dynasty took the famine seriously, but was constrained by more urgent needs to use most of its scarce resources on national defence. This left room for societal and international players. Indeed, this was the first disaster where international relief aid played a significant role in China. As Andrea Janku has shown, western philanthropy, driven by protestant missionaries, triggered also an expansion of local charities, often in competition rather than in cooperation with the Westerners. Many of the local Chinese charities had religious, especially Buddhist, backgrounds and some of them were formed on a cross-provincial basis and became established players in the field. Charities and local philanthropic activities by religious groups and local gentry had already become an established practice by the late Ming dynasty, but now their role in administering relief aid became more pronounced than before.[25] Much of their activities, as Pierre Fuller has argued, took place 'below the radar' of regime gazetteers of foreign observers, which means that their role in late-Qing and Republican relief work might have been much more pronounced than previously realized.[26]

After the fall of the Qing dynasty, the young republic assigned disaster management to the Ministry of Interior in 1912 with its subordinate Department of Civil Affairs taking over the disaster alleviation work.[27] However, the warlord era that followed the virtual collapse of the republic by the end of the decade is usually regarded as a dark period in Chinese disaster management. Funds for disaster relief were often embezzled by warlords and their armies, granaries were looted, food transports for famine victims could be taxed or outright robbed, and river embankments stripped of their protective trees for money, etc. During this time, severe famines occurred in Zhili, Shandong, Shaanxi, Henan, Hubei, taking an estimated 4–6 million lives.[28] Also other kinds of natural disasters riddled the republic. In 1920 a M8.5 earthquake struck what is nowadays Ningxia Hui Autonomous Region claiming the lives of some 200,000 people, and 1931 witnessed extremely severe floods in Central China with an estimated 4 million casualties.[29]

However, as Pierre Fuller[30] has shown, the early warlord period at least did not exhibit any dramatic break from the late Qing period disaster management practices. During the North China Famine in 1920–21, the warlord administrations were 'surprisingly willing, if weak'[31] in providing and even jointly coordinating relief efforts, where foreign relief aid also played a notable role. Local magistrates, charities and individuals, including some of the warlords themselves, were able to organize a relief effort that probably saved the lives of thousands of people. Come the late 1920s, however, cooperation between warlords disappeared, which weakened the 1928–30 Northwestern famine relief efforts notably resulting in 10 million deaths.[32]

The mixed governance system, where local, national and international charities in cooperation with the government, when possible, assumed a major role in relief efforts continued to function throughout the Warlord era. In some ways it even became more entrenched, as with the establishment of the China International Famine Relief Committee (CIFRC) in 1922 on the foundation of pre-existing organizational structures for 1920–2 famine relief. Under weak state conditions, the CIFRC assumed quasi-state functions with the approval of governments in Beijing, and became the most visible single player in the field. However, at the same time reliance on local charities in relief work continued strongly.[33]

When the Nationalist Party, the Kuomintang (KMT), was able to establish an almost unified national government in Nanjing in 1928, disaster management was once more taken under centralizing policies. The newly established central government Relief Office and Relief Commission worked to bring charities and other private organizations working in the field under centralized state control. In practice, however, Nanjing-period disaster relief still relied largely on local actors and charities as far as possible. Even the CIFRC continued its functions until the war against Japan.[34]

The nationalist regime also tried to reaffirm its leading role in disaster relief for reasons of legitimacy. It came to power just when a large famine was gripping Central China and although it did not prioritize relief aid over war and modernization projects, it had to act so as not to appear incompetent. Furthermore, leaving the task to foreigners or private charities fitted ill with the nationalist strong-nation vision. As Andrea Janku has noted, under the new modernizing state, famine and disasters had to belong to the past, and a modern nation had to be able to overcome them on its own.[35]

The official discourse on natural disasters also changed with the regime. Nationalist rhetoric on disasters rejected the old imperial cosmological explanation of the Mandate of Heaven and offered science and technology as solutions to them. In the nationalist discourse, the ability to withstand disasters could also become a patriotic duty of Chinese citizens. During the war against Japan, the Nationalist army used man-made flooding against the invading Japanese forces when the dykes of the Huanghe (Yellow River) were blown up in 1938 near Zhengzhou. The disaster that ensued killed about 900,000 Chinese and made the new route of the Huanghe suffer from constant flooding until 1947, when the break was finally closed. The Japanese were not stopped. Rhetorically, however, the KMT praised the victims for patriotism and sacrifice.[36]

The Maoist Era

Disasters had a connection to the rise of the Chinese Communist Party (CCP) and its ability to defeat the Nationalist regime. However, it would probably be too much to say that natural disasters caused the regime change in 1949. The relation between disasters and the CCP's ability to fight a successful war of revolution was not a straightforward one either. For example, as Odorick Y. K. Wou[37] has shown in Henan, the Yellow River flooding described above created anarchy, dislocation and discontent that the CCP could use to its advantage and flood victims were the main recruitment base for the CCP. However, it was mostly the

Japanese and the KMT's plunder of grain that drove the local population into CCP ranks, when the Communists were able to mount effective resistance to grain raids during local famine. Local CCP leaders would also gain support in organizing effective disaster relief and prevention measures. At the same time, fighting in famine areas caused problems for funding, food supplies and cadre recruitment; soldiers would spend most of their time finding food, not fighting the enemy. Therefore, a famine did not automatically turn into Communist victory in Henan.

Disasters and disaster management also played a part in the CCP's anti-KMT propaganda, but it was only one of the many pressing themes, and not really a horse upon which the Communists rode to victory. When disasters or disaster management were used in propaganda, the idea was to show the superiority of the CCP over the KMT. Already in his speech to the 7th National Congress of the CCP in 1945, Mao Zedong noted how floods, droughts and damage by pests had caused hardship in the fight against Japan, but also how at the same time 'democratic governments', meaning Communist-controlled areas, had achieved unprecedented results in mass campaigns against pests and floods, and how such effective disaster relief had made it possible to continue the War of Resistance. Furthermore, Mao also saw the organization of efficient relief aid for the war refugees and the victims of natural disasters as one of the basic demands the Communists had for the KMT after Japanese defeat.[38]

Right after the Nationalist regime was defeated on the mainland in 1949, unfavourable comparisons of the KMT in relation to the CCP also appeared in communist propaganda in regard to disaster management. The 'old society', as the dynastic and republican eras were now called, also had to be discredited in this respect, and was in many ways certainly a soft target. For example in 1950, Mao would tell the Party Central Committee how in 1949 the country had been hit with large natural calamities over wide areas, but the new people's government had been able to organize extensive relief for the victims, and in many places initiated large-scale water conservancy works. Disaster victims had been given relief grain and now 'things are much better for them [peasants] than in Kuomintang days'.[39]

Notwithstanding the self-praising propaganda, much more of which was forthcoming during the following years, the communist style of disaster management contained both new and old elements. It can be viewed as one of the fields in Chinese contemporary history where the revolution was more like a gradual transition than an abrupt rupture.[40] Continuities started with obvious physical leftovers from previous regimes in the form of trained river systems, flood embankments and reservoirs, but the People's Republic of China (PRC) also copied the general bureaucratic structures of disaster management from the Republic of China (ROC). The overall task of disaster management was assigned to the Ministry of Internal Affairs. In 1950 the Central Disaster Relief Committee was established in Beijing with the task of coordinating the disaster management efforts of the Ministries of Interior, Finance, Agriculture, Water Resources, Health and other related agencies. The committee was abolished in

1958 during the early Great Leap Forward in belief that it was unnecessary in future communist China.[41]

Further continuities could be found in disaster prevention and mitigation practices, such as granaries and principles of relief work, for example, grain-for-work. The CCP also shared the KMT's emphasis on the localities' role in carrying out actual disaster management work, stressing local self-reliance as a virtue. As an inner-party directive put it in 1946, relief work should depend mainly on mutual aid by the masses themselves, and government measures would only complement them. The goal was to encourage the masses to form their own mutual aid teams for disaster governance and mutual self-help.[42] Following these lines, in 1950 the Central Disaster Relief Committee defined the starting points of disaster management as thrift, self-reliance, mutual aid, and using local labour instead of central relief aid in recovery.[43] These principles informed disaster management practices throughout the Mao-era, although they would be manifested differently in different institutional settings as we see below.

But there were also notable changes to disaster governance. The CCP rejected the old late-Qing and republican mixed model of disaster management where the state, private charities and international actors cooperated and competed in disaster management. Charities and philanthropies were either abolished or incorporated into the newly established people's government structure as 'mass organizations', such as the Chinese Red Cross Society, which began to be used for example for local hygiene campaigns. Charity did not fit well with socialist ideology, where the state was seen as responsible, together with the people themselves collectively, for the well-being of citizens, and the maturing of socialism would anyhow make charity redundant.[44] Charities were also often run by religious organizations, which were now being clamped down on and brought under state control, although some non-religious charities could continue working until the Great Leap Forward. Missionary activities were banned and most Western citizens were ejected from the country especially after 'Red China' became a target of Western blockade during the Korean War. Because only American charities really had resources for substantial financial aid after the Second World War, the Korean War dried out foreign charity funding in the People's Republic, signalling the end of foreign charitable work in China that the Civil War and CCP policies had yet already attended to.[45] The Soviet Block welcomed China as its new member, but was not in a position to provide much in terms of disaster assistance, and since the early 1960s, even these relations went sour due to Mao's foreign policies.

In administrative terms, the main responsibility for managing disaster relief work was given in 1949 to the Ministry of Internal Affairs (later the Ministry of Civil Affairs, *Minzhengbu*), and its 'Internal Affairs Section' (*nei wu bu*), which had cadres on provincial, city, county and township levels. The internal affairs administration was responsible for inspecting and reporting on local hazard and disaster conditions, including their size, number of victims, the possible need of grain and other assistance as well as administering relief aid funds. These were decided yearly by the State Council. A large part of disaster management funding was earmarked for emergency aid, which aimed at providing victims with immediate food, clothing, shelter and health care.[46]

However, as a mere section in a ministry, the new disaster governance system did not possess much authority or resources. To illustrate this point, at the end of 1959 when the Great Leap Forward famine was getting severe day-by-day, the Hebei Provincial Bureau for Internal Affairs Welfare Office which was in charge of disaster relief work in the whole province had only 13 staff members, and their number was further reduced in 1961, which was the most serious year of the famine.[47] The office did not therefore have anywhere near the resources needed to cope with the situation – or any other major disaster for that matter. However, many other disaster management functions, such as flood management, were assigned to other ministries and committees as well. The general fragmented nature of the communist party-state was therefore reflected also in the way disaster management was organized during the Mao-era.[48]

Under the conditions of bureaucratic fragmentation and weak professional disaster management authorities, the unifying impulse needed for effective disaster management had to come from elsewhere. In the communist party-state, the overall leadership was always in the hands of the Party. Therefore, it was the CCP, which provided the general leadership for all major disaster management campaigns as well. Only the Party had the authority to make administrative units cooperate during disasters and direct resources to relief efforts from other uses. During disasters, it was the Party, headed by the local first party secretary, which always assumed the overall leadership by creating party leadership small groups and administration headquarters to manage the situation on an *ad hoc* basis. The headquarters then led and coordinated the disaster management efforts of various administrative agencies and was usually abolished after the disaster situation was regarded as having passed. On the whole, this did not necessarily make disaster management weaker, but it became highly dependent on political situations.

In order to better understand disaster governance in Maoist China, the administrative setting is only one of the factors that one needs to be aware of. Indeed, the Maoist state was characterized by a number of socio-political structures which enabled the Communist Party to control and mobilize the population effectively and which therefore also need to be taken into account when discussing Maoist disaster management.[49] Below the local Party Committee, important basic level organizational structures were urban work units (*danweis*), [50] city street committees, and rural people's communes (*renmin gongshe*), which provided the framework for political mobilization during the Maoist era after they were established by 1958.[51] Under them, the most distinctive governance technique used in disaster management was campaigns (*yundong*), or mass mobilization, which many researchers regard as the key to understanding Maoist governance.[52]

Campaigns can be seen as 'concentrated attacks on specific issues through mass mobilization of the populace'.[53] Their goal was usually to solve problems in production and the environment as well as to change the ways people thought about certain issues and social relations. These two goals were usually combined together as campaigns, and especially the bigger ones were combined with an active propaganda effort aimed at explaining to the people the meaning of the campaigns and their role in them.[54] To this end, the ubiquitous role of propaganda in all campaigns is well documented from the early PRC on such as the 'Patriotic Health Campaign',[55] agricultural collectivization,[56] and the campaign to eradicate schistosomiasis,[57] all of which incorporated wider themes of creating patriotism and socialist ethos with aspects of disaster management. Most campaigns were based on popular mass participation, but also more limited bureaucratic campaigns were possible.[58]

Other principles whereby the Maoist society was organized included socio-political segregation of people according to individual class and political backgrounds (*chushen*) into 'red, grey and black' classes, as well as the division of the population into urban and rural residents based on the residential permit (*hukou*) system.[59] Such organizational features were mandated by the needs of the command economy[60] and the ideology of continuing class struggle in socialist society. They deserve to be mentioned here, because all of them played a role in Maoist disaster management decision-making, and especially in the way vulnerabilities became distributed in the society, as we will see.

What this book shows is that these pre-existing organizational structures and practices that were not created for disaster management as such were the key to Maoist preparedness in disaster governance. Indeed, we can talk about a Maoist style of campaign-based disaster management with a high level of general organizational preparedness for disaster prevention and mitigation efforts.

Mao-Era Disasters

With one exception, Maoist era disasters have received much less academic attention than those in the imperial and republican eras. The one exception is of course the great famine caused by the Great Leap Forward (GLF), which has been a topic of a growing body of sophisticated research, so much so that different schemes on how to classify these studies have even been suggested.[61] General historical studies on the period cannot ignore the disaster either and they therefore often describe it in some detail, but typically only as part of their larger narratives of unfolding political struggles in the PRC.[62] However, many scholars have also been interested in the famine itself and especially in the political decision-making both on central and local levels trying to establish various actors' culpability in the disaster from Mao down to village party cadres.[63] This is necessary, as the CCP has basically absolved itself from any guilt for the famine in its 1981 decision on party history by simply noting that the Party Centre had been carried away by the atmosphere of the GLF and therefore made some unfortunate mistakes.[64]

Implicitly, at least, most studies on the GLF famine also address vulnerabilities created by the Maoist state when they try to explain differences in mortality between localities, rural and urban areas as well as those between cadres and common people, genders or age cohorts, etc. Here the discrepancy between urban and rural areas has attracted a lot of attention, but vulnerability differences were also large in terms of localities.[65] Individuals' and local communities' responses and attempts at self-protection during the GLF famine have also been studied, especially in the rural areas. The central role of people saving themselves is argued by Ralph A. Thaxton Jr., who sees the role of government relief aid as miniscule in rural areas and peasants resorting to eating crops on fields as central to their survival.[66] Civil 'resistance' (a CCP term for people and localities protecting themselves) to high state procurement polices during the GLF has also been studied.[67]

However, disaster management during the GLF famine has not yet been studied as systematically, although some studies do touch on the issue.[68] Concerning Tianjin, the role of local leaders' activism for the outcome of the famine has been highlighted by Jeremy Brown, who has shown how city leaders' activities were central to Tianjin grain supplies.[69] Chen Yixian has shown how good leadership mattered in Jiangxi Province, which was saved from the worst famine,[70] and Chris Bramall has made a similar discovery on the importance of local cadre responses in explaining differences in mortality rates between counties in Sichuan.[71]

A common way to frame research questions concerning the GLF famine has been to try to find out how GLF policies caused the famine, not how the famine was related to the general nature of disaster management system and practices in Maoist China. As becomes clear in the following pages, campaign-based disaster management could be an effective way to cope with many types of crises as such, but its major Achilles' heel was its reliance on political authorization from the Party Centre. As long as the centre did not authorize a campaign against a disaster, a city could respond only through ways that did not interfere too much with its other duties. From a disaster management point of view, the tragedy of the GLF famine was that it was a case of two contradicting campaigns, one to mobilize the population to develop productive capabilities, and one needed, but not forthcoming until very late, for mobilizing resources against the causes triggering the disaster. Disaster management campaigns were always disruptions of normal functions of the party-state and command economy, and there were no automatic mechanisms that could trigger an anti-disaster campaign as such in Maoist China. The GLF famine exposed this basic weakness of Maoist disaster management.

Apart from the GLF famine, it is easy to say that natural disasters and especially disaster management are understudied subjects in contemporary Chinese history. Concerning earthquakes, basically only the Great Tangshan earthquake in 1976 has attracted some Western scholarly attention,[72] together with a few single articles on the general Maoist earthquake monitoring system.[73] Even scanter interest has been directed at floods and flood control.[74] This book tries to rectify this situation. The second chapter of the book analyses disaster management during the Great Hebei Flood in 1963. What emerges from these pages is how Maoist disaster management could be highly effective when it had the highest level of political sanction behind it. Indeed, in 1963 Tianjin city rallied after the poor performance during the GLF famine and saved the entire city from a historically unprecedented flood. In 1976 the Great Tangshan earthquake wrecked large parts of Tianjin, but rescue and relief effort was carried out relatively effectively. To be sure, Maoist disaster management had its shortcomings and weak points, too, in these two latter cases, but the Tianjin case does offer an example of how an underdeveloped country could manage disasters relying on its own strengths.

Book Outline

This book deals with famine, flood and earthquake in Tianjin city during the Maoist and early post-Maoist periods. The book's structure also follows these disasters in chronological order: the Great Leap Forward Famine 1959–61, the Great Hebei Flood 1963 and the Tangshan Earthquake in 1976 are analysed in their separate chapters. Indeed, comparatively speaking, Tianjin is a 'disaster city'. No other contemporary metropolitan urban centre has suffered from large natural and man-made disasters at such frequent intervals and variety in China. Unfortunately for Tianjin, this makes it an ideal focus area for an in-depth study of Maoist disaster management. However, such focus also brings obvious limitations with it. This study is, essentially, a case study in Chinese urban disaster management during the Mao era. Although Tianjin had substantial rural areas under its jurisdiction and they also play a visible role in the analysis that will follow, the findings themselves are more urban than rural. It needs to be kept in mind that in the rural areas the disaster management mechanisms described above were notably weaker due to the relative feebleness of the local party-state and the peripheral position of rural areas in power hierarchy and the national economy. However, it is beyond the scope of this book to provide a comparative analysis of Mao-era rural versus urban disaster management as such.

Furthermore, although this book is about Tianjin during the Mao era, it is not a city history as such. Currently, there exists no English language contemporary history of Tianjin, and this book makes no claims to be one. However, to write a book on historical disasters without some reference to a more general city history would be flawed, as would writing a city history without dealing with the natural disasters Tianjin has had to endure. They have been an important part of Tianjin's history and its residents' lives until very recently and shaped the development of the city notably. The contribution of this book, however, lies more in the analysis of Maoist disaster management practices than Tianjin's history as such.

The sources used in this study generally speaking fall into three categories: the archival materials from the Tianjin City Archive (TCA); the internal (*neibu*), publications; and open contemporary sources, especially the gazetteers of different administrative sections and units of the city. Also oral histories, memoirs and media, including online sources, were used to gain a more comprehensive understanding of the events. The local daily, *Tianjin ribao* was analysed as the local party propaganda mouthpiece, and while its reliability as such can often be questioned, in this function it was a very reliable source in saying what the Party wanted it to say. The city archive materials provide mostly first-hand and detailed information on the events and disaster management policies until 1963, and also contain some sensitive information. However, it needs to be noted that the author had access only to the city archives, while the party archives were not open to foreigners at the time of writing. The consequence of this is that the available sources could seldom shed light on the decision-making processes behind disaster management policies, but were reasonably informative on the policies and their outcomes as such. The author feels that this is not a fundamental problem, as the focus of the study is on the general disaster management practices and their outcomes in Mao-era China, but the reader is the final judge of this.

Further, it is good to note that the archival sources from the Great Leap Forward are usually carefully worded, following political currents at the time of their writing, and they most probably leave many things unsaid, especially on the real extent of the GLF famine and food problems in the city and its countryside. Therefore, one cannot regard this information as entirely reliable. The *neibu* publications also served propaganda purposes and therefore had a clear tendency to report only positive achievements. They are, nevertheless, a valuable source for disaster management policies and regulations in general. Contemporary gazetteers are less propagandist and written in a to-the-point manner, but also usually rather sparse on details and circumspect on the negative aspects of the issues. They are a good source for assessing the results of policies and their timings, but one has to remember that, in general, statistical standards, particularly during the GLF, were messy and the accuracy of gazetteers based on such figures can therefore be questionable.

Notes

1 For example Schencking 2006; Oliver-Smith 2002, 24; Hoffman and Oliver-Smith 2002; and Hoffman and Oliver-Smith 1999 argue this way.
2 *The White Paper* 2009, 2.
3 EM-DAT The International Disaster Database. Country Profile: ‘China P Rep’, ‘1900–2009’. http://www.emdat.be/country_profile/index.html (accessed: July 9.7.2009).
4 Janku *et al.* (2012) provide a historical perspective to this.
5 This is argued for example by Edgerton-Tarpley 2014, 251–2; and Li 2007, 5–6.
6 Wisner *et al.* (2004) discuss the relationships between the triggering causes and social process in their well-known PAR model.
7 Bankoff 1999, 394.
8 Quarantelli *et al.* 2007.
9 On vulnerability see Wisner *et al.* 2004, 7–12; Murphy 2009, 3–10; Oliver-Smith 1999, 18–34.
10 Wisner *et al.* 2004, 7–12
11 Boin and ’t Hart 2007, 48.
12 For discussion of these see Schneider and Hwang 2014.
13 Will and Wong 1991, 6.
14 Edgerton-Tarpley 2014.
15 Lewis 2006.
16 Meng 2003.
17 Li 2007, 2.
18 Meng 2003.
19 For the Confucian notion of legitimacy and its intimate connection to disasters see Zhao 2009; Schneider and Hwang 2014; and Edgerton-Tarpley 2014.
20 Rothschild 2012.
21 Elvin 1998, 219.
22 Meng 2003, 114–15, 122–3; Will and Wong 1991, 1–3.

23 'A Comparison between China and Europe' in Li 2007, 5–6.
24 The Qing dynasty's advanced granary system and its function as part of disaster management is analysed in detail in Will and Wong 1991.
25 Janku 2013 and Edgerton-Tarpley 2014 on Qing responses to famines; also Li (2007) has extensive discussion on Qing dynasty and on charities see also Zhou 2006. Shue (2006) and Rogaski (1997) study the role of charities in Tianjin during this time.
26 Fuller 2013.
27 Chen 2014, 30.
28 Fenby 2008, 150–3; a classical highly negative view on the warlord disaster management can be found in Sheridan 1966, 23–4.
29 Chen 2012, 133–4; Chen and Booth 2011, 142.
30 Fuller 2011.
31 Ibid., 54.
32 Fuller 2013.
33 Janku 2013.
34 Ibid., 8–9.
35 Janku 2012 and 2013.
36 Edgerton-Tarpley 2014.
37 Wou 1994, 209, 218–20 and 240–8.
38 Mao Zedong (Tse-tung) 1945a.
39 Mao Zedong (Tse-tung) 1950.
40 Cohen 2003; Esherick 1995.
41 Chen 2012, 134.
42 Mao Zedong (Tse-tung) 1945b.
43 Chen 2012, 134.
44 Zhou 2006.
45 Lieberthal 1980.
46 Meng 2003, 890–1.
47 HBMZHJ 2010, 31.
48 On the fragmented nature of PRC governance see e.g. Lieberthal (1992) and Shue (1988).
49 This draws on the discussion by Nathan (1997); Perry (2007); and Lieberthal (2004).
50 See Bray (2005) on *danweis*.
51 See Perry (2007) on this.
52 Bennett 1976; Strauss (2006); Thornton 2009, 23–48. Perry (2007) sees mass mobilization as the essence of 'Chinese exceptionalism' during the Maoist era.
53 Lieberthal 2004, 65 and 67.
54 Ibid., 65–6.
55 Yang 2004, 155–82.
56 Yu Liu 2006, 732–42 and 737–8.
57 Fan and Lai 2008.
58 Strauss 2006, 897. In his classical study on campaigns Gordon Bennett (1976) offered an alternative typology of seven different types of campaigns, but as he noted, a campaign could belong to multiple categories at the same time and serve both short and long-term needs of the Party.
59 The *hukou* system came officially in force in 1958, but already in 1951 provisional regulations to the same effect were enforced, see Cheng and Selden (1999).
60 See e.g. Naughton (2007) and Bramall (2009) for features of Chinese command economy.
61 See e.g. Bramall (2011) and Chen (2011).
62 See e.g. Lin 2009; MacFarquhar 1983; Teiwes and Sun 1999; and Kung and Lin 2003.
63 Dikötter 2010; Becker 1996; Yang 2009; Li and Yang 2005; Yang 1996; Ashton *et al.* 1984; Bernstein 2006; Wemheuer 2010, 176–94.
64 CCP Central Committee 27.6.1981/1991, 157.

65 Bramall 2011; Chen 2011; Dikötter (2010) has a chapter for example on children during the famine; Friedman *et al.* (1991) argue how peoples' closeness to the party-state influenced their chances of survival during the famine.
66 Thaxton 2008 and 2011.
67 Li 1990.
68 For an exception see e.g. Gao (2011); Li (2007) has a chapter on the Maoist era famine relief in Hebei, but not the larger picture of disaster management as such.
69 Brown 2011 and 2012.
70 Chen 2011.
71 Bramall 2011.
72 Chen 2005; Palmer 2012; and Cheng *et al.* 1988.
73 Bennett 1979; Ross 1984.
74 Ross 1983.

1 The Great Leap Forward Famine

The Great Leap Forward famine of 1959–62 is the deadliest famine in contemporary world history. The severity and duration of the famine varied according to location, but generally speaking it can be said to have begun by the spring of 1959 and continued to the summer of 1961 in most places.[1] Different researchers have estimated the death toll of the famine to be anywhere between 10 and 40 million people, with most researchers agreeing on a number that lies near the higher end of the figures. This chapter analyses how the city of Tianjin responded to the famine.

Before the Great Leap

Tianjin city has historically been one of the most important industrial and commercial cities in China. During the imperial era, it prospered as the terminus of the Great Canal which was the lifeline in the transport of grain from the southern parts of the empire to the capital. As the port city to Beijing, Tianjin (the name means literally 'the Heavenly Ford', heaven meaning a reference to the Son of Heaven, the emperor) became the site of one of the largest foreign settlements in China during the colonial period. During this time, its status and location helped it to grow into a hub of foreign trade and it became a forerunner of modern industrial development in Northern China. It was opened by the Western powers by the Treaty of Beijing in 1860, and, for example, the site of the first railroad built in China. At the beginning of the twentieth century, Tianjin was the number two industrial and sea-port city in China after Shanghai. It produced textiles, light industrial products, metal products, machinery, chemicals, foodstuffs, power generation, and building materials.[2]

During the Republican era, Tianjin was mostly governed by the warlords who ruled Beijing/Peiping. It was granted the status of a municipality in 1927, making it a province-level city. When the War against Japan started in 1937, the city fell quickly to the occupiers and was then ruled under the puppet regime of North China Executive Committee, based in Peiping until the end of the war in 1945. During the war Tianjin suffered from major flooding in 1939, which inundated most of the city proper leaving vivid memories of the ineptitude of the puppet regime in dealing with natural disasters, which the Communists would later utilize in their propaganda.

In the Civil War, Tianjin was liberated on 15 January 1949. The City Party Committee had already been established in December 1948 by the Party Central Committee in anticipation of the city's takeover. At the time of the liberation, the city had 1.86 million people and a land area of 151 km^2. The new administration quickly consolidated its rule establishing the system of people's governments on every level of urban administration, stabilizing prices, and cleansing the 'counter-revolutionary' organizations in the city, but it also provided relief aid to numerous war refugees in the city. The new administration then further strengthened its rule by conducting the anti-three and anti-five campaigns in 1950–2 targetting the members of the old city administration who had remained at their jobs after the takeover and its capitalist classes. By 1952 the city was firmly under Communist command and its economy had recovered to pre-war levels. The first city party conference took place in 1954 and the first City Party Central Committee and the first secretariat were selected in 1955.[3] During all this time, the Tianjin Communist Party was headed by the First Party Secretary Huang Huoqing who was also Mayor until June 1958.

City Preparedness Before the Famine

To understand the starting points of the Great Leap Forward famine management, we must look into the city's food supply system. Here a number of factors were important: the rationing system, separate production and procurement systems for grain and supplementary foodstuffs, and the city's relationship to central grain authorities and the central government in general. Apart from these, the more general Maoist socio-political structures such as the party-state, the *danwei* system, the legal division of the population into urban and rural *hukou* (official residency place) holders, and the mass mobilization techniques discussed in the introduction were of course important factors in their own right. Indeed, they were central to understanding the outcomes in all the three disasters analysed in this study as we will see, but the structure of the food supply system was directly related only to the GLF famine management.

The food supply system was an integral part of the planned economy and therefore its development before the GLF famine followed national policies. The system could be characterized as centralized and monopoly based with some minor exceptions. The first steps in establishing such a system were taken in 1953–4 with central decisions on establishing the monopoly sale and purchase system for grain, edible oils and cotton. By the end of 1954, the Chinese countryside had been placed under state monopoly grain procurement and urban areas under the planned grain supply system. In Tianjin, the City Grain Management Bureau (*Shi Liangshi Guangli Ju*, shortened to 'Grain Bureau' here) was established together with the Central Grain Store to take care of the management of grain supplies. Free grain markets and trading were closed down.[4] Private trading of grain became black market activity. In January 1955, the old yuan became the new Renminbi yuan at a ratio of 1:10,000. During the currency reform, price controls were applied for market goods including foodstuffs. The currency reform was successful, but price controls remained afterwards. Finally, Tianjin followed the national policy of nationalizing private enterprises and the collectivization of agriculture by the end of 1956. This made production decisions state-led, at least in principle, and also moved grain retailing to the state.[5]

The central decisions to collectivize agriculture were also justified by the need for better disaster management. As Mao noted in his speeches and writings in 1955, the way for the majority of the peasants to fight natural calamities was to unite and go forward along the 'high road of socialism', meaning agricultural collectivization.[6] In the book *Socialist Upsurge in China's Countryside* Mao argued that the cooperatives were superior to private agriculture in coping with natural disasters and listed this as one of the main reasons favouring collectivization.[7] Mao did not mention man-made disasters, but with hindsight it is easy to say that the extreme collectivization carried out in 1958 was one of the direct causes of the GLF famine.

In Tianjin the food supply systems could be divided into three segments. The first segment handled staple food supplies (or grains). The second segment consisted of 'supplementary foods' i.e. vegetables, meat, fish, eggs, etc., while the third segment dealt with the rest of what were termed as 'food commodity supplies'. This consisted of food served in restaurants and the 'special supply' system for cadres. Each segment had respective bureaucracy administrating the procurement and distribution activities within its own jurisdiction. Staple food supplies were administered by the Grain Bureau and supplementary foods by the Third Commercial Bureau (later Supplementary Food Office),[8] which also managed the restaurants. As discussed in greater detail below, the constant administrative restructuring that took place during the GLF also affected the way these sectors were organized, but their basic division of labour remained the same throughout the Leap.

The most important commodity for feeding the population was grain, e.g. wheat, rice, millet and corn.[9] A person's grain ration included all these in different combinations and proportions depending on the supply situation. Generally, when the supply situation worsened, 'soft' grain (rice, wheat) would be replaced with 'harder' types (corn, millet, sweet potatoes) and vice versa. Importantly for staple food supplies, Tianjin, and the surrounding Hebei province, was a deficit area. In Tianjin the deficit between the planned procurements and handed-over grain was 3 million kg on average annually, but this of course varied depending on harvests, and as we shall see, the GLF created much larger deficits in local grain production. Deficits were balanced by the central government, which procured grain from all over the country for big cities. For example, in 1957 Tianjin received a little over 701 million kg from outside and produced about 90 million kg in its own suburbs and counties. About 90 per cent of the grain the city needed to feed its residents therefore came from outside including Heilongjiang, Jilin, Liaoning, Anhui, Henan, Hebei, Shandong, Shanxi, Jiangsu, Zhejiang, Jiangxi, Hubei, Hunan and Sichuan provinces at different times.[10]

Importantly for the city's grain supplies, grain was a commodity under the central economic plan. In grain production the city therefore had to rely on central orders and policies, mostly mediated through the provincial authorities when

Tianjin became part of Hebei province in early 1958. However, as a special arrangement, because Hebei could not supply enough grain to Tianjin on its own, the city was still treated like the other big cities directly under central command and remained directly under the central grain supply system even after it was annexed by Hebei province.[11] This opened a direct channel for lobbying the central government during the famine which was of paramount importance for the survival of the city.

In the centralized grain procurement and distribution system, production and consumption areas were dealt with differently. In villages the monopoly purchase and sales system followed the 'three fixed' system: production, procurement, and sales quotas were decided from the top. Crops were divided into grain for food, fodder, and seeds which were left for the producers. How much they got was dependent on the authorities who computed the amount of 'surplus' from the crop. In Tianjin, according to 1955 regulations, households in villages were to be left 384 *jin* (0.5 kg) of unhusked grain per person for a year, and the rest was taxed or procured. Normally 95 per cent of fine grain and 85 per cent of other grain 'surpluses' were procured by the state. These amounts could be reduced or raised based on how good the harvest was. If a locality lacked grain, it could be sold back to it by the state for a higher price than it had been procured,[12] or it could even be given back as relief grain for free, but this seldom happened. Procurement was therefore a form of taxation, which fell heavily on the peasants. During the famine it became a matter of life and death.

In the cities, rationing was an integral part of the staple food supply system. In Tianjin the non-peasant population came under individual grain quotas in November 1955. Officially, this was to guarantee constant food supplies to urban residents, but it had a direct connection to controlling population movements as rations were tied to the *hukou* system, which was also introduced through a number of directives from central government in 1953–8.[13] Basically one could buy rations only in one's own *hukou* area. There were special provincial and national grain tickets as well, but these were not for ordinary *laobaixing* (the common people), but for cadres and other people who had to travel a lot in their work. Of course, a black market existed for rationing tickets, which could also be counterfeited, but resorting to such trade was illegal and required money.

In Tianjin, monthly grain rations were based on the type of work (physical or mental), place of residence and the age of residents.[14] The urban population was divided into four categories with 33 grades. The grades were based on the division of work into light and heavy types. A highest graded worker in the highest category of demanding physical work received 60 *jin* of grain a month, while street residents, who were classified based on their age, received at most 28.5 *jin*. A first-grade brain worker received 32 *jin* a month, while a fourth grader was given 27 *jin*.[15] Cadres could augment their food supplies by using the special supply system with its own shops and also had extra foodstuff subsidies. One of the basic social divisions in Maoist society was therefore the different entitlement to food based on the detailed categorization of the population.

Depending on residents' status, individuals' categories and grades were appraised by *danweis* or streets, which gave these units leverage over improving or reducing their members' food supplies. Officially, in streets the residential small groups were to appraise individual needs, residential committees would report these needs to the street committees, and these would then issue households with Tianjin city resident grain-rationing tickets. Similarly, *danweis* would appraise the needs of their workers and staff members. Rationing tickets were household based, having the name of the head of the household, names of the adult members of the family, work classifications, address and designated grain station written on them. Moving in or out, dying, being sent to a labour camp (*laogai*), being born or any events resulting in changes in the number of household members were to be reported promptly to keep records straight. Minority groups and foreigners had their own regulations. Muslims were for example supplied with mutton instead of pork.[16] The bottom line of the system was that urban residents did not have any clear-cut and objective right to certain amounts of food, but how much grain they got was in the end decided in a political process. The way the appraisals were made became a central issue during the famine.

Grain was *the* staple food, but the diet of common urban residents also consisted of what was termed as 'supplementary' (*fushi*), or non-staple, food where vegetables dominated. Vegetable supply was actually greater in weight (but not in calories) than that of grain. Importantly, as perishable goods with lots of local variation, supplementary foods were a commodity group under the city's own economic planning, which meant that special interest was directed at them as supplements to, and sometimes also as substitutes for, grain. Tianjin city began to plan vegetable production in 1956 because of growth in demand due to city population growth. Before this vegetable growing and sales had been free. On a grassroots level, supplementary foods were distributed through district-level retail companies, which managed the street-level retail shops and markets.[17] Industry and commerce received grain based on economic plans and certificates.[18]

Vegetables for the city were mostly grown in the near suburbs and during the GLF they were assigned the duty of 'serving the city' with vegetables. Outer counties were assigned the task of grain growing and producing vegetables for their own needs. This was mainly based on pragmatic reasons of transportation difficulties for large and heavy loads of beets and roots to the city.

An important factor influencing both grain and vegetable supplies was the crop cycle. In Hebei, there is the spring grain harvest in June and the autumn grain harvest in September–October. The flour proportion of grain rations followed the harvests and fluctuated between 10 per cent and 60 per cent of the full ration. The rest was made up with what was available, often tubers, such as sweet potatoes, but when they were in short supply, other parts of vegetables such as the tops, petals and leaves of plants were counted as 'grain'. In practice, during the GLF famine people were mostly given anything but good quality grain flour in their 'grain' rations. The supply of rice also fluctuated. Between 1959 and 1961, residents were sold on average 1 *jin* of rice per month but, at low points of supply, they did not receive any real rice at all, only its residues and millet.[19] On average, urban residents' rations varied quite a lot from season to season even before the Great Leap Forward. In 1955 the average ration was only 26.32 *jin* per person, but then rising to 29.32 *jin* in 1956. In 1957 it was once more time to save grain due to a bad harvest and in March 1958 an average grain ration was 28.57 *jin*.[20]

The picture was similar with vegetables, which were in short supply in January–March and then improved gradually with early varieties ripening from late March. March–May were dominated by leafy vegetables, such as celery, then until August the fruit types, such as peppers and tomatoes, and then from August to October tubers, leaf and stem vegetables were in balance. In November the *baicai* (Chinese cabbage) crop entered the shops in such great numbers that it had to be piled up on street corners for customers to take home from there. It would then be stored throughout the winter forming the main winter vegetable, although it was liable to rot.[21] Vegetable rations normally fluctuated between 2–5 *liang*[22] a day, but official targets could be as high as 7–9 *liang*. Both grain and vegetable planting and harvest seasons set the pulse of mass mobilization not only in the countryside but also in the city, where many urbanites, such as students, were recruited to help during the busy seasons. However, during the GLF famine the participation of large parts of the population in planting and harvest campaigns became extremely urgent and finally a priority for everybody.

The crop cycle was also affected by natural conditions. The Hebei plains, where Tianjin is located, is monsoon area. Typically there can be long periods of drought ending in heavy rains. In the Tianjin region, a winter–spring drought was the rule, not an exception. Rains would usually come in July–August and when they came, they could create waterlogging or even flooding. Both drought and flooding were therefore something that could be anticipated and guarded against and anti-drought mobilization was as normal a feature of the agricultural work cycle as planting and harvesting campaigns. The food crisis gave them new urgency, however.

The third category of food supplies was the products of the food processing industry. This sector was rather prominent in Tianjin and managing it was also under the Third Commercial Bureau. The food-processing industry had 'six big products': cakes, fried cakes (*shaobing*), *mantou*, *mifan*, noodles and bread. All these were subject to ticket rationing. Products were standardized for size and quality and had fixed rations by which they were computed back to grain. For some minor products such as *baozi* and *jiaozi*, biscuits and gruel no rationing was imposed at first, but sales were limited to the amount a person could eat on the spot as no take-away service was available.[23]

A notable part of supplementary food was distributed through markets. After the nationalization of the industrial economy and the collectivization of agriculture in 1956, the Party Centre decided to still allow 'free markets under state leadership' as a supplement to state markets. The rationale for this was to improve market variety and circulation and thus keep industrial workers happy. In April 1957, markets in Tianjin were reorganized so that all city districts got a total of 31 agricultural markets, and city districts established Market Offices for supervising these. In addition to this, in the same year more than 60 small peddler 'synthetic (i.e. joint) markets' were operating under city management. They sold some two-thirds of all agricultural by-products by value in the city.[24] It was estimated that some 4,000–6,000 peasants came daily to sell their goods in Tianjin from the suburbs, outer counties and Hebei province, and some even from other provinces like Shandong and Liaoning. They especially sold supplementary food, which was the most common product at the markets. Each day some 500,000 kg of vegetables and 50,000 kg of seafood were sold. Other market products included fruits and peasant handicrafts.[25] In September–November 1957, a stricter market control system was established by the State Government for different industries from vegetables to bicycles. It included ordering retailers and peddlers to buy their goods at authorized wholesalers, not directly from suburban producers, and vegetable prices became more strictly regulated.[26]

This monopolistic system was challenged from the beginning both by individuals and work units that engaged in bypassing the regulations in various ways. Most commonly, they would go to procure their supplementary foodstuffs directly from the suburban producers. Sometimes this was motivated by the inability of the centralized system to deliver enough good-quality foodstuffs for the *danwei* or individual household kitchens, sometimes by profit-seeking. Both phenomena increased notably during the GLF crisis, but already before the GLF began, bringing individual economic initiative under control was a major concern for the Market Management Bureau. Apart from work units and individuals going directly to the villages to buy vegetables, peasants would come to the city. It was estimated that in 1957 there were still some 7,000 small peddlers without a licence in the city.[27] In early GLF the city targeted these peddlers with a campaign pressuring them to form cooperatives.[28]

When the GLF began, one had little legal choices of gaining access to alternative food supplies in the city apart from state outlets. It was also evident that both suppliers and consumers broke the regulations when possible and this could only grow more serious as the crisis developed. The control of food distribution therefore became a battleground during the GLF.

An Administrative Mess

Other, man-made, factors influenced the outcome of the famine in Tianjin. The administrative status, organizations and boundaries of Tianjin were under constant fluctuation during the GLF, so much so that it can be called an administrative mess. In the spring of 1958, Tianjin city lost its status as a municipality directly under the central government and became a part of Hebei province. This was decided in February by the State Government and has been seen as the result of the lobbying of Hebei's provincial Party Secretary Lin Tie who saw Tianjin as a key to the economic development of Hebei.[29] However, the reorganization also meant that Tianjin became the administrative capital of the province instead of Baoding City that had served in this function earlier.[30] The merger took place in April 1958. As a consequence, the city's position weakened in central bureaucracy and among provinces. Fewer resources from central government were forthcoming to the city than before. Because of this, Tianjin is said to have suffered more in economic terms than other big cities from the economic policies of the late 1950s and 1960s, and the city stayed behind other big cities in economic development as long as it was part of Hebei.[31] Nevertheless, it was still an important industrial centre so the loss of status did not immediately translate into lesser demands for production during the Leap.

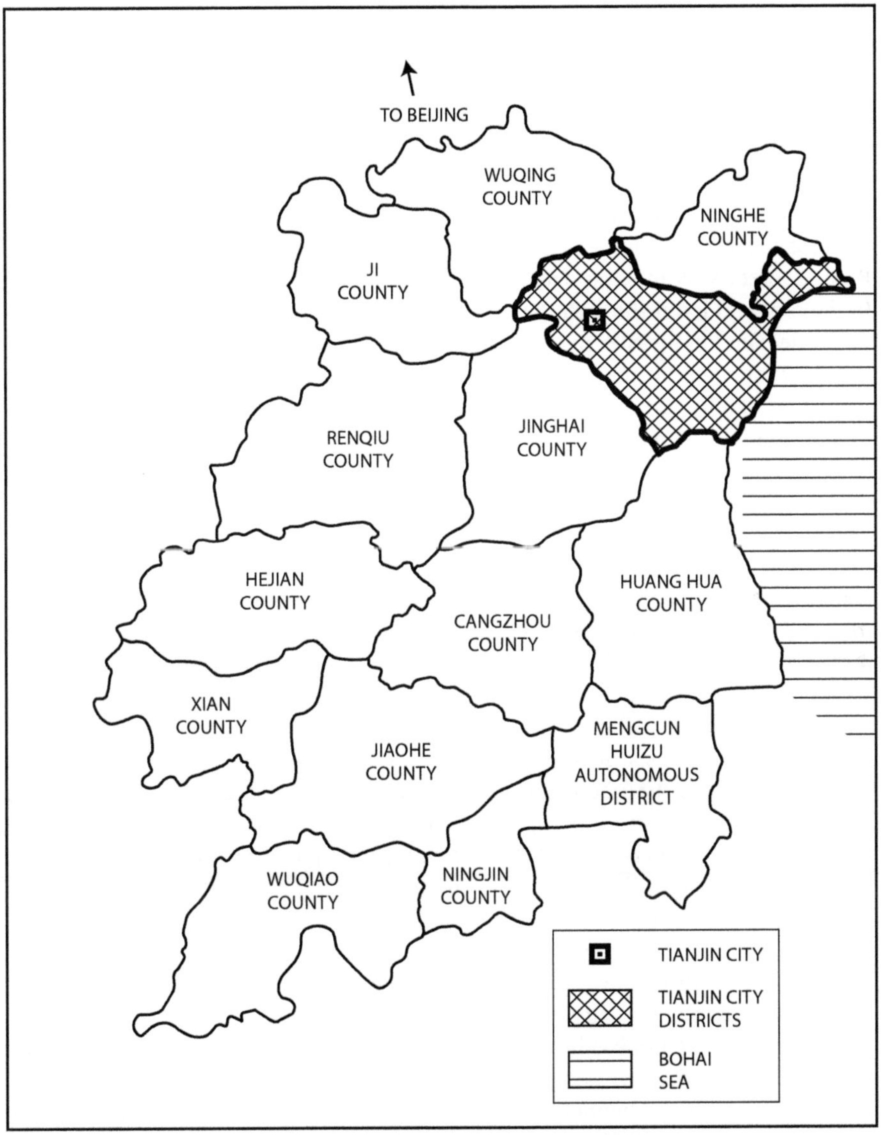

Map 1.1 Tianjin's administrative borders and division at the end of 1959

After this, a major reorganization of the city's organizational chart took place almost every six months. Bureaus and offices responsible for feeding the city were also subjected to these upheavals. First the administrative system was reorganized in the spirit of central instructions on decentralization of economic management when the central government released its controls of its own enterprises to localities in June 1958. This also happened in Tianjin, where control of most of the city companies, including those responsible for grain distribution, was delegated to city districts. Such delegation of powers was typical of the early Leap period, but it was soon deemed dysfunctional, which became evident in Tianjin already in February 1959, when the city began to recentralize its grain supply administration.[32]

When the city became subordinated to Hebei province, City Party Committee leaders also changed. In June, Tianjin Party Secretary Huang Huoqing was transferred to Liaoning, and Vice-Mayor Wan Xiaotang, a cadre with a poor peasant background who had previously worked for the police administration, was nominated as the Tianjin First Party Secretary and Li Gengtao became Mayor.[33] It was Party Secretary Wan, who declared the grand goals for the Great Leap Forward in the city and led it through the GLF famine. He was also personally behind all major decision in disaster management during this time, but in this he was careful to follow the central line closely.

As importantly, the area of jurisdiction of Tianjin city also fluctuated constantly. During the GLF, a number of counties and even one prefecture were first merged and then separated from Tianjin. This began in February 1958, when the Hebei Provincial Government decided to move Hangu and some other counties to Tianjin.[34] Then in October 1958, Tianjin reorganized its urban districts. The original four urban districts and six suburban ones became eight city districts: Heping, Hedong, Hebei, Nankai, Hexi, Hongqiao, Hangu and Tangu.[35] Even bigger change took place in December 1958 when the Provincial Party Committee decided to merge Tianjin and its prefecture counties[36] (12 plus one transferred from Tangshan) with the city. Now the city area had grown from 2,533 km^2 to 20,000 km^2, with 11.4 million people within its borders stretching from the Shangdong to Beijing.[37]

Jeremy Brown has argued that Tianjin city leaders were ill-equipped to handle rural affairs suddenly assigned to them through these mergers and overwhelmed with the problems caused by the rural food crisis when it began.[38] The mergers indeed may have over-strained their resources, which led to the relative neglect of rural areas also in Tianjin disaster management. The eight southern counties which the city absorbed in the merger were regarded as particularly poor, producing little, if any, surplus grain. Already before the merger there had been food riots in Tianjin Prefecture in the spring of 1958 and some peasants from these counties had ended up begging on the streets of Tianjin.[39] These counties would become the worst famine-hit areas in the city and therefore, in effect, the city inherited a large disaster zone through the merger.

In 1960, three more counties were transferred to Tianjin from Tangshan expanding the city area to 34,000 km^2. Reversing these mergers began in July 1961 when Tianjin Prefecture was re-established and placed under the dual jurisdiction of the city and province.[40] Six months later, the prefecture was transferred directly to provincial jurisdiction. All these organizational changes were accompanied with reorganization of the Tianjin administrative structures as well.[41] The situation where one reorganization followed another messed up the administration and weakened its ability to respond to the crisis effectively. For example, the Grain Bureau annals note that after 1958 in GLF the grain and edible oil management system was in constant flux weakening its management capacity.[42]

The Descent into Crisis Early 1958 – Summer 1960

Disaster management in Tianjin followed the course of the Great Leap Forward in three distinct phases: from early 1958 to late summer 1960 the city responded to the worsening food supply crisis through temporary measures aimed at making its food reserves last until the hoped-for improvement of the situation. Then in the late summer of 1960, the tightening grain situation and central decisions to cut urban rations also brought widespread hunger to the urban areas of Tianjin and the city entered an emergency mobilization phase that lasted until the following summer. After this, a gradual recovery phase began lasting until the end of 1962.

The spring of 1958 was a spring of high hopes and aspirations everywhere in China. In late summer 1957 the Communist Party had declared ambitious agricultural development aims. After this, central party leaders gave similar targets to industry and all other realms of life during the winter and spring of 1958.[43] According to Chairman Mao, and the party leadership that toed his line, China would overtake the industrialized world within a few years (first 15, then only 3 years). China would begin transformation into a truly communist society, the first of its kind in the world. To achieve this utopian target, the Party mobilized the masses on all fronts to exert their outmost for economic development. An extraordinary mass campaign that was to be called the Great Leap Forward began. So it was also in Tianjin, which was swept over by an avalanche of new ambitious projects, production targets and campaigns during the spring of 1958.

All sectors of society were mobilized to carry out the leap in the city. In *Tianjin ribao* the 'Great Leap Forward' was first mentioned in mid-February when the paper called for more support for the leap in agriculture.[44] Industry was also caught up in the high tide. The 2nd Five-Year Plan for the city was revised and industrial investments in the city went up ten times compared to the 1st Five-Year Plan aiming at reaching the level of Shanghai in production within five years. This meant the tripling of industrial output. For 1958, the city leadership decided that the industrial gross output was to grow by 45 per cent and individual industry sectors were assigned even higher targets.[45] In addition, the face of the city was to be made over. During the Leap, the city initiated over 400 major urban construction projects, most of them in 1958. For Tianjin, this was almost solely an urban leap, as 99–99.5 per cemt of fixed investments went to urban districts.[46]

All sectors of the urban society were mobilized to carry out the leap. In mid-January the City Government made a decision on mobilizing a campaign against the 'four pests'. The city was to become 'the city without the four' (i.e. rats, flies, locusts and sparrows).[47] Following the decision, in the first week of February the city mobilized 1 million residents to eradicate the 'four pests' and Mayor Huang inspected the results. This was only the first wave in the pest eradication campaign; on 1 April, the City Party Committee and government declared that the four pests had to be eradicated before the May Day and two other mass mobilizations were carried out in April. For example, on 6 April 1.9 million residents were mobilized in a day-long campaign against sparrows. By May Day the campaign had claimed the lives of 3.85 million sparrows.[48] In May the Grain Bureau announced that its granaries had 'basically' achieved the goals of the 'ten no's' policy it had announced at the beginning of the year: no mildew, worms, rats, sparrows, mosquitoes, flies, work-eaten grain, rubbish, accidents or weeds were to be found in the city granaries that were leaping forward with the rest of urban society.[49]

Tianjin also followed all other national initiatives, including initiating massive water works and a steel production campaign. From 20–25 January 1958, the city convened large agricultural, forestry and water conservancy project meetings, where the mobilization of the countryside for the GLF in agriculture for 1958 was announced. In the spirit of the ongoing anti-right campaign that had begun in June 1957 after the short-lived 'Hundred Flowers' campaign, it was emphasized that 'rightist deviation thinking', i.e. too timid attitudes towards agricultural growth, had to be overcome. In this spirit, new kinds of experimental fields and high-yield plots were also promoted in the Tianjin countryside.[50] The subsequent failure of such fields was one of the reasons for the food crisis that started to develop in late 1958.[51]

Large water works were also carried out in Tianjin. In late June 1958, the City Party Committee issued an order to the whole party and residents to participate in the Redoing Haihe Project under the slogan of 'clearing the river by redirecting waste waters [away from Haihe] and separating salt and fresh water from each other'. The aim was to build a sluice at the Haihe river month so that seawater could not enter the river and reconstruct water channels in the city in order to prevent waste waters from entering the river thus making the river water clean and clear. In early July, city leaders called for mobilizing 'all people' for the project.[52] This was not the only water works project in the city, however. In the Tianjin countryside in 1958 communes constructed 196 reservoirs, and attended 11 different large water construction projects claiming to have achieved irrigation of 70 per cent of Tianjin fields.[53] Such large water works projects demanded a lot of workers, who came almost solely from the countryside. Combined with the steel-making campaign in autumn 1958, it led to severe shortages of agricultural manpower at the time of grain and vegetable harvests. The consequences were clear to see already in the winter of 1959.

On some fronts Tianjin even tried to take the lead nationally, such as establishing urban communes, a number of them in the summer of 1958.[54] In mid-July the Hebei District reported it had established an urban commune with a joint eatery, kindergarten, washing and sewing stations, library, and a residents' club. Many visitors visited the place. By August, six urban communes were established in

Tianjin and were noted nationally. As in the rural people's communes, urban communes were also to take care of production, goods allocation including food distribution, exchange, and welfare of their members, combining industrial work, agriculture, commerce, study and military functions. They were to become the new form of the socialist city.[55] Newly constructed residential buildings were planned without household kitchens. Even individual household toilets were abandoned and collective latrines constructed instead. This added to general collectivism in the GLF, but also to inconvenience for residents.[56] Maybe it was for the better that building residential buildings was not a priority during the GLF.

As local history notes, after the Central Committee Beidaihe meeting in August 1958, where the leaders of the CCP decided on drastic expansion of the GLF, Tianjin's economic planning was transformed into constant 'leap planning' with targets changing all the time. The City Government for example now decided that the Redoing Haihe project had to be completed within three months, thereby shortening the project time by half.[57] It was decided to carry out communization of agriculture in the city's countryside by the end of August. Communes were given the task of 'arduous struggle for three months, large-scale water works, implementing deep ploughing and dense planting, liberating women to work by establishing kindergartens, communal mess halls, and collective sewing services.'[58] Four trial communes in the Hongqiao District and some other areas declared that they would achieve communism in 3–6 years.[59] Local history notes that now the 'leftist' mistakes of bloating production figures, the 'wind of communism', commandism, etc. started to spread seriously.[60]

With the communes came a new way to distribute food and salaries. By October 'basically all' peasants in suburban districts were eating in large public mess halls.[61] Following the directives from the provincial Party Secretary Lin Tie, communes set up a 'half free supply – half work point compensation' system, whereby half the peasants' compensation was given as 'free' food in the public mess halls, where peasants were practically obliged to eat. In Tianjin this was accomplished by November when 2,345 mess halls had been established in communes in city suburbs.[62] People were told to eat their bellies full.[63] The significance of this system was that it restricted legally the peasants' entitlements to food to mess halls and thus made controlling individual peasants easier during the food crisis that was about to start. Numerous bitter local disputes would arise because of the way mess halls were run.

Like the rest of the country, Tianjin industry and agriculture were launching satellites (*weisheng*) time and again.[64] On National Day, *Tianjin ribao* reported how in Xinli Village of the Eastern suburb, an experimental field had produced over 124,000 *jin* of grain per *mou*.[65] The village had been inspected by Mao just before the Beidaihe meeting. Such results were bogus, but they did feed the fever of bloating production figures. The satellite fields also contributed to the crop failure that followed, as everything planted in them died. When combined with the policy of transferring labour to water works and urban industries and leaving an increasing amount of acreage fallow, grain production fell dangerously all over China. However, with the expected high crops agricultural taxation also grew tighter: in 1958, the surplus procurement rate in the new people's communes was 95 per cent.[66]

The Beidaihe meeting also initiated a massive steel-making campaign, with problematic outcomes for the city's food supplies. An important part of the GLF was the attempt to develop steel production to the levels matching and even surpassing the UK and USA. Tianjin was one of the national steel industry bases to begin with, producing about half of Hebei's annual steel products and little less than 5 per cent of the national total.[67] Following the decisions made in the Beidaihe meeting, the provincial Party Committee decided on the target of producing 105 *wan* tonnes of steel by the end of the year – about half of which was Tianjin's share.[68] The Beidaihe decision on steel production was announced to the public on 1 September by *Tianjin ribao* together with other decisions on establishing people's communes, mobilizing large water works, implementing trials on deep ploughing and other 'scientific' land-improving methods and the continued campaign against the four pests.[69] On the same day city steel workers held a meeting where they pledged to reach the year's new target.[70]

In order to reach this goal, units were ordered to use 'local methods' (*tufa*) to supplement the modern steel sector. By mid-October, 3,880 small backyard furnaces were constructed in the city to support the steel campaign.[71] Two million people, including peasants in suburbs and counties, were made to live and eat on furnace sites with no rest or chance to go home.[72] The city also decided to construct four new steel factories by the end of the year. In the chaos of these rush projects, hundreds of people were going around construction sites not knowing what to do, and public mess halls did not even know how many people they were supposed to feed.[73] Furnaces produced inferior steel and left massive amounts of waste around them. Trees were felled in the city parks in order to feed the furnaces turning many public places into virtual dumping grounds.[74] But the campaign had direct consequences for food production and distribution as well.

The Food Crisis Begins

All these campaigns laid the basis for the food crisis in China and Tianjin. In urban Tianjin, the early food crisis was a by-product of the nationwide steel production campaign. The city leaders announced an 'eat full' policy for the nearly 2 million people participating in the steel campaign and raised their grain rations from on average of 29 *jin* to 31 *jin* a month until March 1959. The campaign also took away labour from the Tianjin countryside and transport capacity from grain shipments and, as a consequence, Tianjin received only some 59 per cent of the required grain in the fourth quarter of 1958. By the end of 1958, the city was depleting its granaries.[75] Moreover, because the campaign also removed labour from vegetable growing, the autumn crop was collected poorly. About 40 per cent of *baicai* crops, which was about 25 per cent of the city's vegetable supplies, perished in the fields.[76] In total, city agricultural production declined by 7.1 per cent in 1958.[77]

Although the press was filled with official propaganda about the extraordinary accomplishments of experimental fields and other projects which led to the impression of abundance in making, the reality for the city food supplies and other basic necessities was already rather bleak in late 1958. At the end of 1958, an internal report from Tianjin countryside indicated how many of the villages lacked grain or could not get it in timely fashion, the food supply system was called 'chaotic' and people were eating so many seeds that about one-third of seeds for winter wheat planting were missing in five Tianjin counties.[78]

These problems became even more pressing during the spring of 1959. By the end of 1958, common residents in the city could already also see clearly that all was not right in the economy. A report on the city market situation to the provincial Party Committee at the end of the year noted how there were shortages of supplementary foods, such as vegetables, meat, shrimps, seafood, and pastry. Some daily necessities were also in short supply, such as soda powder, sugar, firewood, soap, napkins, batteries, enamelware, *baijiu* (strong Chinese liquor) and tobacco paper. Agriculture and light industry were suffering from many shortages, such as leather, paper, chemicals and fuels.[79] The steel production campaign created other hardships for residents, too. In mid-October, the City Government issued an order to further economize on coal consumption, which meant less coal for private heating and cooking in order to reserve more of it for steel production.[80] In mid-December, the City Government issued an order to economize on electricity for the same reason.[81]

The quality of food rations began to deteriorate by this time in the city. At the end of December, the City Government cut the wheat flour proportion of rations from 30 per cent to 20 per cent, while the food industry got a 50 per cent cut. Wheat was to be substituted with rice and sweet potato flour.[82] However, although city leaders were already aware of mounting problems in supplying the city and its residents, their faith in the Leap did not falter in public. At a city party meeting at the end of 1958, City Party Secretary Wan still proclaimed the need to make an even bigger leap forward in 1959.[83] The mass mobilization of the population for different production campaigns continued.

Some cause for jubilation was given at the end of the year when it was announced that the Haihe Sluice and the city waste-water projects had been completed on schedule, and Lin Tie and Wan Xiaotang also attended the celebrations.[84] The wastewater project was actually still continuing, but the construction of the great sea sluice was finished, and it was the largest water construction project in Tianjin history. Normally it would have taken three years to complete, but now it had been finished in just six months. It was seen as the symbol of GLF achievements in Tianjin.[85] Later on, the sluice did indeed serve some useful purpose in controlling tidal waves in the Haihe river system.

However, there were fewer reasons for celebrations in the city economy otherwise. The city's industrial and construction sectors were facing serious shortages. Factory stoppages and production slowdowns were common, and the general level of production began to fall in the beginning of March 1959 continuing throughout the rest of the GLF.[86] The city leadership was well aware of these problems and they were addressed on 18 January 1959, when the City Party Committee convened an enlarged meeting of all party cadres. Party Secretary Wan declared that although the city had reached all its production goals for 1958, the goals for 1959 were facing difficulties because raw material supplies were not enough. In the first quarter, apart from cotton, raw materials demanded by the city's industry could be met only at the rate of 10–30 per cent, at best 50 per cent.[87] Wan noted that this would lead to shortages of daily industrial products. He blamed the tense market situation on the increased population in the city, but also criticized the system of free food in the villages as unstable. It had led to 'not a little' waste and also deprived the city of vegetables that it needed. As Wan put it, this all meant that 'we have to be mentally prepared for long-term shortages'. Therefore, he urged cadres to mobilize people to 'eat a little less', leading cadres too had to lower their own living standards on their own initiative, and resist developing any special privileges, Wan demanded.[88]

By February 1959, city grain reserves were at their lowest since the beginning of the monopoly grain system. In February, the city received only 24.18 per cent of its required grain. Reserves in city granaries dwindled to 21 million kg.[89] The City Government reported at the end of January that after the end of 1958 the supply of supplementary foodstuff had suffered from shortages and that vegetables were often sold out in the city markets.[90] The spring crop was also a disappointment due to, *inter alia*, spring drought, and procurement fell notably short of its target.[91]

Nevertheless, central pressure for meeting procurement targets was high in the late winter of 1959. Although the spring of 1959 is seen as a period of relative relaxation of GLF goals by Mao, in grain procurement the opposite was true. Industrial production targets were cut somewhat, but not grain procurement targets. In February, Mao approved a campaign against concealing grain, which he saw as the reason for the discrepancy between reported production and actually procured grain figures. The campaign gave local cadres the authority, if they needed any extra, to search and confiscate all grain in households. Mao also raised procurement targets for the spring of 1959.[92]

Under these conditions, the food crisis worsened in the Tianjin countryside. This was related especially to the poor condition of the newly acquired southern counties. Already in the winter of 1959, some counties were sending reports to the City Government on food problems. For example, in early February, Wuqing County reported that it lacked 27.5 million kg of grain, and 20 million kg of fodder and fuel to survive the winter.[93] Later, the same county would become infamous for its famine-related mass poisonings and irregular deaths.

By June the situation led to widespread hunger in Tianjin countryside, which was revealed in a report to the city's Party Committee about an oedema epidemic that had developed in Huanghua, Jinghai, Cangxian, Jiaohe, Yanshan and Wujiao counties during the winter and spring. Reportedly, the epidemic had started in April and the counties had all together 8,446 oedema patients. People were suffering from swollen bellies and limbs, some becoming so swollen that they could not see. Seventy per cent of patients were women; often those breastfeeding their babies, it was reported. The epidemic had hit both villages with mess halls and those where people ate at home. The report further indicated that the main cause of the epidemic was the lack of protein in the food people were eating. In April many areas had cut their grain rations to 12 *liang* a day, 60 per cent of which was given in sweet potatoes, which were poor in protein. Further, a large part of the rations was wasted in the process of preparing food, so the 12 *liang* ration was real only on paper. In the areas south of the Tianjin-Shanghai railroad, people ate a lot of *xianhuangxucai*, which was a highly alkaline wild vegetable, negatively affecting digestion when eaten a lot. It created stomach problems like diarrhoea in many villages and also caused poisonings. In Xian County, which was also affected by the epidemic, people had eaten poisonous wild vegetables and rotten potatoes with the result of 128 poisoning cases and three deaths.[94] However, another report on the same epidemic gave a figure of 16 dead from eating poisonous wild vegetables.[95]

The report further blamed the lack of hygiene standards in mess halls for the problems. Some mess halls did not even wash wild vegetables before preparing them. Salt, vinegar or garlic were used too little so they could not sanitize the vegetables either, it was complained. Eating mouldy foodstuff was also a problem.[96] The report claimed that a number of measures had been taken to stop the epidemic. County leaders had held a telephone meeting discussing the situation, and the Health Bureau had inspected problem areas and taken preventive medical (both western and local) measures, sending teams of medics to villages. The principle was to 'cure those with money and those without it'. The City Health Bureau had sent seven doctors to Huanghua, Jinghai and Cang counties to help. Medicine included vitamins B and C and fruit sugar, as well as chilli pepper and garlic water. Other measures included improving mess hall management and vegetable treatment hygiene, as well as educating people on hygiene. Jiaohe, Renqiu and Cang counties had printed small leaflets on the reasons of oedema and distributed them to villages and held meetings to propagate proper ways to mix wild vegetables with food.[97]

In some counties the most severely affected villages were given relief grain to improve the flour proportions of peasants' rations. For example, in Huanghua County, Luqiao Commune, extra wheat flours had made it possible to lower sweet potato ration from 70 per cent to 50 per cent. Potatoes and garlic were also shipped to villages.[98] What the report did not mention was that while these measures were being taken, Hebei province was still demanding that the counties deliver their share of grain production to Tianjin under the slogan of 'guarantee Tianjin grain supplies'.[99]

The report was very representative of the disaster reporting to the higher levels by health authorities during the GLF famine in China. As Yang Jisheng has noted, such reports never mentioned starvation directly, but instead talked about 'epidemics' thereby framing the crisis as a medical, not food-distribution related issue.[100] Also the report from Tianjin counties was carefully worded not to implicate the ongoing production campaigns on the problems. It did indicate that the problems that existed were caused by the lack of protein, but never discussed the high procurement rates which caused the shortages in the first place. The report also clearly downplayed the severity of the problems and, at least implicitly, put the blame for the health problems on the famine victims themselves for eating the wrong kinds of herbs and preparing them in unhygienic ways. The lack of protein in food was linked to the wild vegetables and sweet potatoes, which were both promoted by the authorities themselves as grain substitutes. The figures for victims given in the report also appear very low. According to a contemporary memoir, Wan Xiaotang received a report about the same epidemic indicating that around 840,000 people were living in the affected disaster areas, and that 1,270 villages had serious problems with food.[101] It is extremely unlikely that only about 1 per cent of the population in these counties would have been affected by edema. Such understatement was probably due to both falsified local statistics, the political sensitivity of the issue and the fact that authorities clearly wanted to frame it as a small to medium-sized medical problem and nothing else. But such understated reports were followed by understated interventions.

It is not a surprise that Wan Xiaotang and other leaders resorted to going on inspection tours of the countryside to receive more accurate information throughout the GLF. For example, in the summer of 1959, Vice Mayor Niu Yong and municipal Grain Bureau Director Liu Pichang visited Cang County and were appalled by what they saw there, even after the intervention described above. People were getting only watery gruel, vegetable leaves and stalks to eat. The leaders decided to 'lend' food to the counties without approval from higher levels and were later censored by Premier Zhou Enlai for that.[102]

These events say a lot about how the party-state managed the food crisis in the countryside. The criteria whereby it decided who was eligible for relief aid and who was not were not discussed in the reports. What is clear, however, is that the City Government framed the problems as medical emergencies and therefore favored short-term medical interventions whereby the city and counties organized teams of medics and shipped in some relief aid and medicine to selected localities, after which the situation was declared as basically solved and the problem, if not as disappeared, then at least as being under control. Of course, this could be partly explained by the way leaders expected the next crop to solve the supply problems, but this hope never materialized during the GLF. Instead, in Tianjin at least, the inefficient medical interventions developed into a repeated pattern.

The City Saves Grain

While the Tianjin countryside was already facing an oedema epidemic and 'unnatural deaths', as the official euphemism called famine deaths, city residents were also feeling the pinch in their rations in the winter of 1959. The situation of low grain reserves prompted the city leaders to launch a campaign to 'economize on grain consumption' in February after Wan Xiaotang's speech discussed above. Although it was originally meant as a temporary measure, the campaign continued more or less relentlessly throughout the rest of the GLF. As noted above, rationing as such was not a new thing in Tianjin. However, in the grain-saving campaign, rationing was combined with active redistribution of urban residents' private food reserves, which was new. As part of the campaign, the extra rations for steel campaign workers were terminated in March. Then, at the end of May, the City Party Committee issued orders on cutting rations in urban areas, which was based on the Party Centre's decision. According to the orders, grain consumption was to be reduced to the level of January last year by cutting approximately 1.5 million *jin* per year of grain from residents, 1.67–2.5 million *jin* from temporary residents, 1.4 million *jin* from suburban peasants and 1.4 million *jin* from alcohol factories. Food processing industries' and fodder rations were also cut somewhat.[103]

The order was typical in that cuts were divided over the whole spectrum of consumers and industries and that people without a permanent *hukou* and suburban peasants were made to take a notable share of the cuts. However, the official order was only part of the picture. The city had already started grain saving in February based on 'voluntary' cuts in individual rations. Households were expected to hand in their 'surplus' grain to the city and accept, and even actively suggest, cuts to their individual rations. In effect, this doubled the reductions of average rations and some families were subject to much larger confiscations of grain from their private reserves.[104]

District Grain Bureau offices and units under them were put in charge of this campaign. They were to take back grain from residents through propaganda work, visiting households and persuading them to cut their rations, but also to conduct house searches in order to find 'surplus' grain. Districts could proudly report success in getting households to cut their rations by more than the City Party Committee had ordered.[105] In the Hongqiao District the Grain Bureau managed to get people to hand in 798,805 *jin* of grain until late May. It propagated the grain-saving policy, selected model workers in its own workshops to promote it, praised model households that saved grain, and made *danweis* clear their warehouses of food reserves. It had also used the 'three cores and one correct' (*sanhe yidui*) guideline to rectify people's rations: checking their *hukous*, grain cards, and accounts and verifying whether people's rations corresponded to their real situation. In the process, the bureau had discovered 18,399 households, which had not reported that their members had moved out, died or been arrested, and were therefore still eating their grain. Some were discovered falsely claiming that they had lost their food tickets and were thereby getting double rations.[106]

The district party committee organized 11 'promoting grain-savings' action groups that studied policies and promoted them by visiting households after work. The groups could be persistent in applying pressure on residents. Reportedly a district resident named Zhang Hongli and his six-member family were not willing to cut their 340 *jin* monthly rations, claiming that all family members required 60 *jin* [sic] a month, but after three visits by an action group to this family, they confessed that they just wanted to eat white bread and therefore their rations were so high. They cut their own rations reportedly voluntarily by 118 *jin*.[107]

The Heping District Grain Bureau conducted a similar household rectification campaign removing the dead, moved-out and arrested from its rosters. Propaganda and education were also carried out to make residents hand over their grain voluntarily. In April, district activists mobilized residents to cut a total of 448,168 *jin* from their rations. The average monthly consumption went from 29.61 to 28.97 *jin* in the district, which was more than the city had expected. Moreover, food-processing *danweis* in the district handed back 78,216 *jin* of grain to the city.[108]

The grain-saving campaign utilized the whole urban administrative structure to reach the residents' private food reserves. As a report on the campaign by Hexi District Huidui Street from October 1959 demonstrated, grass roots authorities engaged in the work actively. In this case, a number of people in the street had complained about 'grain shortages' [*sic*, in quotation marks in original]. As a response, the street party committee had formed a special Office for Rectifying Grain Supplies with 14 cadres from the street party committee secretariat, the neighbourhood committee, police station, finance and commerce committee cadres, the street Youth League, the Women's Federation, the local grain station, and a representative of the district Grain Company. The residential committees in the street also established small groups to handle the work. Propaganda work directed at the residents was carried out and 14 large residential meetings were held to discuss and propagate the policy.[109] However, the idea was not to increase grain supplies and thus solve the complaints, but to show how people themselves were demanding too much and were therefore to blame for the situation.

The Office for Rectifying Grain Supplies conducted a survey of the situation in the street and its report explained how residents themselves were behind the circumstances: sometimes people used grain in an unplanned way, such as by keeping excessively large reserves. They could even be feeding their chickens with their grain or exchanging grain tickets for clothing tickets. Based on the survey, it was discovered that one-quarter of households actually received too much grain and only 5 per cent received too little. The one-quarter with excessively large rations had to cut them and return their surplus grain to the city. Allegedly, this not only solved the grain shortage problems in the street, but each month's supplies of grain could be cut by 670 *jin* and the state also got some 4,900 *jin* of surplus grain back.[110]

The Maoist state relied on peer pressure to get people involved in its campaigns and this was the case in the grain-saving campaign, too. A good example of this was given by Xiandezhuang Street in Hexi District, where 48 households consisting of 228 people were reported to have established 'residents grain-usage mutual aid associations' in June 1959 under the street office to promote grain saving in the street. Also in Xiandezhuang some households were reported as having received too large rations, while 19 households had 'felt' [*sic*] that they had too little rations. When the street office proposed the formation of the association, some surplus households did not want to join, while the deficit households welcomed the idea. To persuade the surplus households to join, street cadres helped activists in the street to educate and propagate correct ways of thinking, stressing how saving grain was important for the economic construction of the country. Surplus households were further told that they should show 'friendly socialist spirit' of mutual aid to help other people; while households in deficit were criticized for their willingness to rely on outside help and told to plan their grain consumption more carefully. The party-state was therefore placing the blame for the situation equally both on those with too much and too little grain.[111]

In the end the residents made a common agreement on saving grain that everybody in the association was expected to supervise. Surplus households were to hand in their leftover ration coupons 'voluntarily' to the association at the end of each month. The association also promoted the planned use of grain, but also a number of methods to make food last longer, which became propagated by city authorities during 1959. These included food augmentation methods, which usually meant adding water to food, saving grain by mixing it with vegetables; and eating breakfast a little 'dry' and supper a little 'wet'. Following this advice, one 30 *jin* household was reportedly turned into a surplus household.[112]

The definition of 'solving' grain shortages therefore had a very different meaning depending on whether you asked the authorities or street residents. 'Rectifying' grain supplies in a street usually meant that people lost part of their rations and even their private reserves. These reports showed how the city felt it appropriate and within its jurisdiction to confiscate grain from city residents, not only peasants when needed, although the former were handled much more softly. Food was not a private matter. Typically for the Maoist state, people were made to take part 'voluntarily' in a campaign that could harm their own interests. The street offices' tactics to pit poorer households against richer ones was also evident. Such tactics of divide and rule were typical for Maoist political campaigns.

The campaign also involved industrial units. In this, the City No. 1 Steel Factory was offered as a model. For the purpose of economizing on grain consumption, the Factory Party Committee had organized the Factory Staff Living Conditions Committee, and all workshops had organized small groups for the same purpose. The factory also established a 'quantity supply card' (*dingliang gongying kapian*) system, whereby staff members' grain usage was planned for ten days ahead, in one-month intervals. At the end of the month, all accounts were settled. At the same time, staff members were given political education on economizing on grain consumption through posters, notice boards, cartoons and meetings. As a result, the factory could boast that it had saved 187,000 *jin* of grain from March to December 1959 and handed over 140,000 *jin* of grain to the state. The main saving method was 'augmenting' its foodstuff. It was claimed that this method not only saved grain, but also produced food that was so cheap and good that workers 'competed to buy it'.[113] Water did not contain any calories even in Maoist China, but it did add to the weight of rations. As food shortages continued, the quality of rations deteriorated further. Urbanites were therefore more poorly supplied than looking at official rations would lead one to believe, yet always better than the peasants.

'Blocking the Leaks'

Apart from forcing urban residents to save grain and edible oils, the city began to block 'leaks' in the grain supply system more vigorously during the spring of 1959. This policy concentrated on two items. First, Tianjin leaders tried to restrict entitlements to food to Tianjin urban residents alone. Second, they also tried to curb individuals and *danweis* from breaking regulations and independently arranging their food supplies.

In April 1959, while starvation was spreading in the countryside, the city authorities limited the amount of food people could export from Tianjin. Organizations could export food only if they had valid certificates for it, and individuals could carry outside the city only small amounts of food as snacks. The Post Office was ordered not to carry shipments containing food.[114] A bureaucratic campaign was launched to enforce these regulations. The Grain Bureau sent inspection teams operating with the police to catch people smuggling grain out of the city. Following practices that had already been used before the GLF food crisis,[115] 26 checkpoints were placed around the city to check outgoing people for contraband. People used their creativity to hide foodstuff; it could be hidden inside clothing in coffins ostensibly for burials. Women could be faking pregnancy to carry foodstuff, or hide it inside fake babies carried by their 'mothers'. Much of the smuggling was blamed on career criminals, who were identified in one-third of the cases. They were running operations where some members of the groups lived in Tianjin and were connected to *neibu* personnel in units handling foodstuff. After trading with these corrupted cadres, other members of the groups smuggled food away from the city and a third group of people then sold it further. Some bought non-rationed foodstuff such as cakes in great quantities and transported them outside. Sometimes all products were sold to these dealers, and city residents were left without. About one-fifth of the people caught transporting foodstuff were common workers, especially temporary construction workers, staff members or cadres. They usually carried 20–30 *jin* of food with them and it was often unclear whether it was for their own consumption, bringing large gifts to relatives or for trade. The most common markets for foodstuff were in nearby counties.[116] The fact that such smuggling assuaged starvation in the villages did not stop city authorities from trying to curb the trade.

However, city authorities tried to control the movements of food both ways and therefore tried to curb also urban individuals and work units from taking food supply matters into their own hands. A major problem from the point of view of city authorities was that many *danweis* circumvented the official supply chains and went directly to food producers in villages. Time and again the city had to issue directives banning such activities. This began already in November 1958, but it is clear that *danweis* did not follow the directive, because the ban had to be reaffirmed in April, June and August 1959[117] and then again in 1960 and 1961.

Already in November 1958, the City Government had banned *danweis* from going directly to production brigades to purchase vegetables there. The directive reminded people that vegetables should be sold only to district supplementary food retail companies, but especially units with comparatively large kitchens had hooked up directly with communes. The directive named the City First Cotton Linen Factory and Teachers' College as bad examples of such behaviour, which was making a mess of the city economic plan and breaking market order.[118] In January, the Bulletin complained how some units were still circumventing the new system of purchase certificates on vegetables, pork and mutton, complaining how some *danweis* were trying to 'superficially' improve the lives of their members by sending their staff members to the suburbs to purchase *baicai*, pigs, peanuts and dates. The Bulletin also complained how cadres in *danweis* were taking part of the vegetables supplied to cafeterias for themselves, which was creating a bad atmosphere when vegetables were often sold out in the markets.[119]

The policy of making communes sell their products to the state monopoly according to plan and banning *danweis* from buying these products directly was once again reaffirmed in Market Price Regulations issued in April based on provincial decisions. These were also issued in order to curb inflation in the situation where some retail shops were 'wantonly' raising their prices over list prices and criminal elements and black marketers were engaging in speculative activities, it was claimed. The city now ordered prices to be cut to the level of the third quarter of 1958.[120] Inflation was indeed rife. Depending on the month, springtime vegetable prices in the city went up from 35 per cent to 81 per cent compared to 1958.[121] In general such directives had little effect, as they just drove trade under the counter with higher prices.

The weakness of such regulations was shown in the way the City Government had to again order city units to immediately stop the practice of going on their own to the suburbs to buy foodstuff at higher prices in June 1959. One needed a good reason and permission from the Third Commerce Bureau if one were to buy directly from outside. The City Government also reminded food-processing industry units that they were not allowed to sell their produce (such as meat, fish, sugar and cakes) on their own, but all had to be sold to the state. Citing difficulties in getting raw materials could not be used as an excuse for not supplying city markets. The order threatened the infringing *danweis* with investigation and 'rectification'.[122] In July *danweis* were further banned from renting vehicles from villages to transport their goods for lower cost, as this practice made it more difficult to transport agricultural products to the city.[123]

The attempts to curb outflows of food and unauthorized shopping trips by *danweis* to the countryside brought forth the earliest form of GLF famine propaganda in the spring and summer of 1959. In it, both the *Tianjin ribao* and the City Bulletin attributed the reason for food shortages in the city to individuals' and units' deviant, illegal and politically incorrect activities, which messed up the food supply system. These same units and individuals were also said to have been the ones that complained about the lack of food. Shifting the blame for the food problem on individuals and deviant units was an attempt to hide the party-state's role in the creation of shortages. When the famine evolved, other explanations were added to this line of argument, but this one was never abandoned. Famine propaganda was therefore organic in its nature, and explanations grew and evolved as if constructing a protective canopy around the hard core of the reasons for famine: party-state policies and mistakes.

As part of crisis management, reducing urban population was also included on the political agenda in the spring of 1959. The City Party Committee had ordered a halt to hiring new people for urban industries before the Central Committee meeting on 18 January 1959, which decided to moderate GLF targets. By this time, over 200,000 people had already been hired in the city, mostly from the suburbs and villages.[124] The City Planning Bureau complained in April that the rural population that had been transferred to urban jobs and that had migrated 'blindly' to the city had constituted in 1958 13 per cent of the rural workforce in some of the city's counties. In 1959, this inflow had just continued. The Planning Bureau demanded that city work units cease recruiting peasants wantonly and follow centrally approved plans. Further, illegally hired people should be sent back. This was especially the case with those with bad class background who were to be deported 'without exception', while ordinary peasants without permanent employment should be 'generally' deported home.[125]

Managing the numbers of blind migrants, in effect often famine refugees, henceforth became one of the key issues in Tianjin disaster management. As a part of the grain-saving campaign in April district grain bureaus began a campaign to check the *hukous*, grain accounts and how people's rations corresponded to their real situations to make sure that people were not getting undue rations.[126] Such 'checking and verification' campaigns especially targetted people without an urban *hukou* or work contracts. Checking and verification became a constant feature of disaster management in the city, but this was not only about having fewer mouths to feed. Reducing the urban population was also connected to the promotion of food production by the city itself, which gradually became more and more important.

Promoting Food Production

Promoting food production by urban residents was connected to the crop cycles described above. In February 1959, the city mobilized urban residents to collect farmyard manure for the suburban areas to improve vegetable production as part of the spring planting campaign.[127] In March 1959, urban work units were told to participate in an oil plant growing campaign by using idle lands, such as roadsides and ditch embankments, to grow castor-oil plants and sunflowers. However, unlike peasants, urbanites were not given production quotas. Further, the motivation of this campaign was not only to improve nutrition, but also to produce export earnings. The campaign directive noted how 1 tonne of oil could be traded for over 1 tonne of steel products. To provide further incentive, all communes that sold 100 *jin* of oil seeds to the state were entitled to buy 5 *jin* of edible oils from it. Urban *danweis* were allowed to keep 20 per cent of sunflower seeds that they produced.[128]

The campaign did not work. In July 1959, Hebei province decided to stop altogether the supplies of edible oils to villages and cut sales in Tianjin by one-third. Residents' monthly edible oil ratios went down to 6 *liang*, and intellectual workers, administrative staff and students lost most of their extra rations, but received still 4 *liang* more. The city also supplied extra edible oil rations to heavy workers and people working in high temperatures.[129] Dearth was also felt in other ways. At the end of June, the Hebei government cut clothing rations by 60 per cent. People were told to use supplementary materials instead of cotton for clothing. *Danweis* were no longer delivered cloth even for propaganda banderols.[130]

City authorities also paid close attention to the summer harvest. In late May 1959, the City Party Committee and Government issued instructions on summer harvest arrangements urging people to do their outmost for the harvest and replanting. All should have been completed in ten days to two weeks. Harvest was to be precise, well conducted and clean (clearing all). Urban workunits (schools, regiments, industrial companies, etc.) were also mobilized to help in the harvest. Peasants were told to gather every granule and reject erroneous ways of thinking, such as 'the loss of some grain in bumper harvest does not matter'. Quick harvesting and getting grain into granaries was emphasized. Not a single granule was to go missing.[131]

In mid-June, the Party Committee and City Government issued further instructions on the summer harvest noting how the grain supply situation was tight at the moment and how therefore the planned amounts had to be produced and purchased to the full. The central policy of 'quick harvest, quick threshing, purchase, and transportation' had to be carried out thoroughly. It was further noted how under the 'national grain difficulties', which were now admitted at least in a *neibu* City Bulletin, that surplus brigades could keep up to 80 *jin* of grain per person (for summer, i.e. until the autumn harvest in September – October). Brigades producing over 60 *jin*, but less than 80 *jin*, were to settle for what they had as the state would not resell grain to them.[132] Implicitly, this made 60 *jin* a limit at which the state would intervene by allowing the reselling of grain.

As noted, spring and summer of 1959 was a period when the Party Centre moderated its most utopian goals for the GLF. Mao, who had been made aware of the various problems and even outright famine conditions that the Leap had already caused in some parts of the country,[133] visited Tianjin once again on 23 February making remarks on the need to rectifying people's communes.[134] Party Secretary Wan relayed Mao's remarks to Tianjin cadres after hearing Hebei's report on its situation during the visit.[135] This period also marked slightly more relaxed propaganda work, relatively speaking. As seen above, when the summer harvest season began in June 1959, the City Party Committee and Government could momentarily admit to the tight grain supply situation and even that there were 'national grain difficulties', but this was not to admit to any failures by the Party. Instead it was to motivate people for a rush harvesting campaign. According to the official version, the reason for the tight grain situation was that the last autumn's crop had been 'lost, rotten, spoiled, and used in unplanned ways, as well as waste and damaging of the crops that took place a lot'. This time people had to be taught how to harvest grain carefully and diligently and welcome this year's higher purchase quotas 'dancingly, vigorously and eagerly'.[136] Here the blame for shortages pointed directly once again to peasants and local cadres with their sloppy harvesting, storage and transportation work as well as 'unplanned' use of grain by urban people. The theme of using grain and vegetables in a 'planned manner' became central to propaganda by summer 1959.

While the city was by and large dependant on outside shipments of grain, it could do something about its vegetable supplies. As noted in the beginning, vegetables belonged to the city's own economic planning and its suburbs were assigned the duty of growing them solely for urban consumption. When grain got scarce, vegetables supplemented it. During spring 1959, urban residents were getting a meagre 2–3 *liang* of vegetables a day, but by June spring crops improved vegetable supplies to an average of 6 *liang* a day, although this was still well below the official 1.4 *jin* standard (for peasants 1 *jin*).[137]

Under these circumstances, city authorities tried to improve vegetable production. In June, Tianjin representatives took part in the '14 cities meeting' held in Beijing. The meeting laid out the new central policy of making 'vegetables the key' to the suburbs. According to this policy, suburban areas were to improve vegetable supplies to their central cities by 40 per cent by the end of the year. In Tianjin, city leaders seized this opportunity to improve urban food supplies. The City Party Committee and Government relayed plans on increasing vegetable production in June at a meeting of responsible cadres for vegetable production. In the meeting, Vice-Mayor Song Jingyi, who would play a key part in food supply matters in the city during the famine, noted how shortages in supplementary food supply were causing popular resentment. It was therefore decided to expand vegetable growing acreage and improve storage work.[138] The workforce for this was to come from sending back to the suburbs the 80,000 surplus temporary workers from the city and village industries that lacked the raw materials to be run properly. Land to cultivate was there: in the suburbs only 79 per cent of the fields reserved for vegetables were actually cultivated and in the counties the figure was even worse, only 49 per cent.[139]

Promoting vegetable production was stepped up when the autumn planting season approached. In early July, the City Party Committee launched a '1,000 *jin* grain, 10,000 *jin* potatoes, 100 *jin* ginned cotton [per *mou*]' bumper harvest competition for the autumn harvest season.[140] In the same month, the Central Committee made a decision to make localities responsible for promoting their own supplementary foodstuff production and ask for central assistance only as a last resort when lacking grain.[141] In Tianjin Vice-Mayor Song Jingyi relayed this call for *zili-gengsheng* in food production at a meeting of 1,300 responsible cadres and government workers from all levels on 23 July.[142] The message was that the city had to rely on itself in supplementary food production and that city leaders were taking this opportunity to improve food supplies seriously. This can be said to be the beginning of the vegetable production campaign that now became another permanent feature of city disaster management.

The End of Moderation

Moderation in the GLF, however, proved to be short-lived as the general political atmosphere changed drastically soon after this. At the Lushan conference in August 1959, Mao clashed with Defence Minister Peng Dehuai with the outcome that the GLF entered another high tide when Mao rekindled its zeal in a response to what he termed 'rightist opportunism'. Calling for moderation and criticizing any part of the GLF became political heresy and on 7 August the Party Centre issued a directive on 'Opposing rightist ideological deviation'. In Tianjin the City

Party Committee convened an emergency meeting to study the directive and on 14 August Party Secretary Wan gave a speech entitled 'Raise High the Banner of General Line, Resist Rightist Thinking, Quickly Raise another High Tide for Production and Economizing' at a large cadre meeting.[143] He ordered industrial and agricultural units to inspect themselves for rightists and a hefty 31 per cent general growth target in value was set for production in 1961.[144] Still relatively new to the office, Party Secretary Wan thus showed his ability to toe the central policy line closely.

The new political atmosphere could also be seen in how propaganda became more assertive in the issue of food and other shortages. On 20 September, *Tianjin ribao* reprinted a *Renmin ribao* article attacking head-on complaints about market shortages as 'slander'. The article claimed that national markets were 'flourishing'. Daily commodities supplies were growing, storages were full and prices were stable. The state had provided relief aid to victims in disaster areas. Complaints about shortages were thus 'totally groundless'. However, because China was a large country and cities' purchasing power was constantly rising, some temporary shortages and 'relative shortages' in the markets could occur as long as production forces had not yet had time to catch up with demand, the article also explained.[145] *Tianjin ribao* added its own commentary to the *Renmin ribao* column noting how the market situation in the city was also 'flourishing'. Market supplies had improved quantitatively speaking, notably belying the 'right conservative slander' about shortages. This said, it was also noted that in order to improve market supplies further, people had to get rid of their rightist deviationist thinking and the production and economizing campaign was stepped up in the city.[146]

The articles presented the typical double-talk that was becoming increasingly used in famine propaganda by the summer of 1959. The 'relative shortages' explanation gave shortages a positive spin by attributing them to rising living standards that it was claimed people were enjoying and became the basic explanation for grain and vegetable shortages throughout the famine. It was a way to tell people that shortages were actually a sign of success, and that economic development was just so quick that supplies could not keep pace with demand – people had less because they actually had more!

A close relative to this argument was the argument for economizing more in order to produce more. According to this explanation, markets were flourishing all over China, but people had to economize on their spending and work more enthusiastically in order to further improve supplies and avert the crisis that, officially, was not there. Such spin on shortages was for example given by Vice-Mayor Guo Chunyuan who gave a speech in late July on the ongoing Economizing and Increasing Production campaign to the City People's Congress meeting, arguing that the constant savings that people were urged to do were actually to make the economy produce more. According to Guo, people should understand that economizing on consumption was about making it possible to produce more and thus actually about improving people's living standards.[147] Following this spin, consuming less was actually consuming more!

Economizing on consumption also meant further grain rationing. In mid-August, the Central Committee ordered a new campaign on economizing and increasing production, which was also touted in the *Tianjin ribao*. The decision demanded that all cities and the countryside had to conduct a savings education campaign concerning grain, coal and those commodities that as yet were 'not amply supplied'.[148] In Tianjin grain-saving targets for the year were increased from 98.5 million kg to 102 million kg. Different sectors in the city were assigned their own saving targets: rural areas 47 million kg and urban areas about 55 million kg, of which about 22 million kg were assigned to the food-processing industry, 26 million kg to households (both as cuts to rations and as voluntary returns to the state) and 7 million kg to temporary workers. About half the urban quota had already been saved in January–July.[149]

In the ensuing grain-saving drive, urban residents ended up saving over 10 per cent more grain in the third quarter of 1959 than had been planned.[150] As an example, the City Government bulletin offered to emulate a man who had undergone self-criticism for not being able to hand in all agreed 'surplus' grain to his residential committee.[151] The situation was made more tense by the fact that, in the last quarter, the city population grew by 190,000 people, but grain consumption went down by 1 million *jin*.[152] This alarmed city leaders so much that in late November 1959 the City Party Committee issued a directive on curbing the hiring of new personnel for work units.[153] In September 1959, the city also forbade non-*hukou* people from buying any foodstuffs in city markets.[154] At no point were people told the total saving targets.

The more radical atmosphere also influenced the way the city handled refugees. In October 1959, the city began a campaign to better control what was termed 'blind migration' and established the Office for Dissuading the Agricultural Workforce from Blindly Taking Refuge, later named the Office for Repatriation of the Blindly Migrating People.[155] The name was well chosen, as sending 'workforce', not refugees, back to its places of origin was easier to justify even under the newly radicalized conditions. A large migrant reception station ('Tianjin Shourong and Repatriation Station'), in effect a refugee camp, was erected at Tianjin East Railway Station and all city districts established their own *shourong* ('reception and handling') offices. The railway station used to be the favourite place for inflowing vagrants to live before the GLF,[156] so the choice of the place for the refugee camp was not a coincidence. In 1958–61, 600,000 blind migrants were handled through this system. The Tianjin Shourong and Repatriation Station guidelines defined its target customers as: 'those outsider peasants who come to the city without board and lodging, and those urban people who roam the streets without means of living'.[157]

Initially, those refugees who could work were organized into labour teams to earn their tickets home. In 1959, 110,000 refugees were made to work on construction sites. The Office for Dissuading Agricultural Workforce from Blindly Taking Refuge also established relations with the provinces and counties of origin of refugees. Hebei, Shandong and Anhui were the most prominent ones of these. Many of them sent special teams to Tianjin to get their people back and

Table 1.1 The number of repatriated 'beggars' (cases) in Tianjin 1957–63

Year	*Number*
1957	2 009
1958	116 229
1959	148 296
1960	227 587
1961	110 283
1962	43 003
1963	26 510

Source: MZHJZH 1999, 299

Tianjin also dispatched its personnel to see refugees back to their home villages. Some 'habitual drifters' (*xi liu*) were processed many times over, as they returned to the city once repatriated.[158]

The Tianjin countryside also got its share of the newly radicalized atmosphere. Villages were targetted with another rush grain procurement campaign during the autumn harvest. The City Party Committee and Government instructions on harvest noted how units that were complaining that too much grain had been procured from them had to undergo inspections of their real situation. Those who complained for nothing had to be subjected to political education to resolve their 'thought problems'. Real deficit areas were to get their grain according to state regulations, but all areas, whether on surplus or on deficit, had to continue economizing grain consumption under the slogan 'being frugal during rich years' and promote eating more vegetables instead.[159] The Grain Bureau gave instructions on how using 'wild vegetables' should be expanded in the villages, and how eating them mixed with grains should be promoted. Wild vegetables was a euphemism referring to anything somehow edible growing in nature, such as grass. As part of the autumn harvest, 270,000 villagers were now mobilized to collect wild vegetables to then hand in to mess halls as substitute for grain.[160]

On 23 October, the City Party Committee convened a meeting of county and district-level financial secretaries where Vice-Mayor Song Jingyi demanded that the procurement movement on grain, cotton and edible oils be sped up so that the procurement targets could be completed in advance.[161] Officially, procurement was indeed completed successfully by mid-November one month before the deadline.[162] It was announced as a great victory, but the reality was different. The 1959 grain procurement target was *c.* 108.5 million kg, but actual procurement was only 60.5 million kg.[163]

It is clear that rush procurement was unpopular in the countryside. Basically, rush procurement was one way to prevent peasants from eating the crops themselves. Therefore, it directly contributed to severe lack of food in villages and did not even help the city much, as rushed harvest and threshing just created more waste, often intentionally.[164] More than a few granules were lost in the process. The city leaders were aware of the situation and used the anti-rightist opportunism campaign to carry out an investigation into the reasons for poor procurement results.

At the end of December, the City Party Committee received a report from the Countryside Commune Rectification Office on carrying out the anti-rightist opportunism campaign in the Tianjin countryside, addressing the 'unhealthy trends' (*waifeng*) in grain procurement.

The 'unhealthy trends' did not mean excessive procurement by the state, but peasants hiding grain and complaining about the procurement policy. The report noted that many localities had developed unhealthy trends and blamed peasants for the grain shortages. Peasants had lots of complaints, it was reported. They were complaining that the state was 'demanding too much [grain]' from them, '[grain] production was deficient', '[grain] production quotas were too high, so they could not be met', 'the state was … squeezing grain from peasants', 'all the work force for the year had already been given to the state', and that 'when leaping forward, no one was managing to feed the people, and there was no way to arrange proper living conditions for them'.[165]

The report also noted how peasants everywhere were hiding grain and dividing it up in secrecy. For example, all 17 communes in the Wuqiao County had done so. Their methods of hiding grain were many: handing in wetter grain, accepting damaged grain as good, and reporting more people and cattle to feed than there actually were in the villages. They could also report lower increases in production and more in cuts to them. Also reporting more on disaster conditions and wasted land, and less on increases to acreage were used to hide true production figures. Peasants were hiding grain in different places like 'black warehouses', inside house walls, beneath rice straws, buried underground, in the wilderness, and even in fake graves.[166] In the Hebei countryside, a village rectification campaign that focused on concealing grain from the state continued throughout the spring of 1960 making it more difficult for peasants to sustain a living.[167]

With all these problems, the *Tianjin ribao* claimed that there had been yet another bumper harvest in the autumn both in vegetables and grain.[168] Sticking to this claim, propaganda was, as before, trying to blame shortages on ordinary people. By the end of 1959, the City Bulletin told its urban readers at the same time how food supplies were better than ever and that they had to learn to eat vegetable peels properly in order not to waste the precious nutrition in them.[169]

The tense political atmosphere affected supplementary foodstuff production in a contradictory manner. The June plans to send part of the non-city-*hukou* population back to suburbs for increasing vegetable production there fell through as this labour was needed in the city to continue the industrial leap forward.[170] However, promoting supplementary food production was at the same time stepped up – and not without some tragicomic features. In October, Chairman Mao expressed his support for the Hebei province Wujiaowang Xiansi People's Commune, which had begun to promote raising pigs in a big way. This was now made a model to emulate for the rest of the country. In Tianjin, Vice-Mayor Song Jingyi took charge of the campaign to promote pig rearing in the city. Based on the two central policies of promoting vegetable growing and pig rearing, the City Bulletin announced in November that the goal was to attain self-sufficiency or a surplus situation in all supplementary food products by 1962. This meant increasing vegetable cultivation by 2½ times and making all work units rear their own pigs.[171] *Tianjin ribao* echoed the aim of resisting rightist conservatism and realizing the target of 'one person one pig'.[172] According to official figures, over half the work units in the city were made to engage in vegetable production and pig rearing during the winter and spring of 1959–60. Vegetable-growing acreage was expanded by 36 per cent and the number of pigs in the city increased by over 100 per cent to 2 million head. However, the plan was to have 15 million pigs in the city by 1962.[173]

The pig-rearing policy in particular was not popular as work units lacked resources and experience to carry it out properly, but criticizing this ill-conceived policy was regarded as showing a 'rightist' attitude. To promote the policy, the City Bulletin ran articles introducing production brigades that had been successful in pig rearing, such as a Jinghai production brigade where commune leaders had reportedly promoted it with great results. Not only had the brigade stopped living on state relief as it had done before but, it was reported, the brigade could repay its loans to the state and its funds improved thanks to the policy. Villagers could buy bicycles, radios, new houses, and build a new school.[174] This was all empty propaganda, as even official history notes that promoting pig rearing led to losses due to being too large scale with low expertise and fodder problems.[175]

In December, the city leaders assembled all leading cadres responsible for food production at a five-day meeting at Tianjinshi Renmin Hotel – all invited brought their own ration tickets – and decided to use the *zaliang* (food grains other than wheat and rice) production areas for corn and sweet potatoes in 1960, as they yielded the best results. In vegetables the most emphasis was put on *baicai*.[176] At the end of the month, the City Government established the City Supplementary Foodstuff Committee to strengthen the leadership and management of supplementary food production and sales. The committee was headed by Song Jingyi.[177]

By the end of 1959 the *Tianjin ribao* was touting bumper harvests in all fields of agricultural production in the city. Grain production was said to have increased by 9.8 per cent, agricultural supplementary production by 19 per cent, acreage yield by 28.7 per cent, vegetables by 37 per cent, pig head count by 80 per cent, and seafood by 36 per cent.[178] In reality, none of these measures prevented the food situation from deteriorating throughout the winter of 1960.

Hitting Rock Bottom: 1960

1960 began with discouraging signs. In mid-January 1960, heavy snow fell on the city. Right after this, the City Government convened a meeting aimed at getting the transport sector to pledge to save fuel. The coal-saving campaign was also stepped up during the winter and spring.[179] Although Tianjin was covered in white, 1960 was a black year for famine in China. After the Lushan conference, Mao did not receive, or did not act upon, any reports of famine for six months,[180] and even after this it took another six months until any substantial changes in central policy were made to deal with the situation. Although the city leaders were well aware of the shortages of basically everything but manpower (and even that was in the wrong places), on the surface the city continued the GLF as if nothing was wrong. Indeed, the fact that the highest Tianjin leadership had executed GLF-related policy initiatives vigorously enough was shown in the fact that no leading figures in the city were removed from their posts during the anti-rightist campaign. Ever-higher production targets were assigned to different sectors. At the end of February 1960, the City Party Committee decided on the production targets for industry and agriculture in 1960, setting an unrealistic 30.91 per cent as the general growth target for value.[181]

Early February brought some relief to urbanites when the Chinese New Year improved supplies momentarily. The city had prepared for the festivities and ordered food-related *xitongs* (administrative units) to prepare well for it so that there would be a good variety of fresh products and sufficient amounts to 'reflect the prosperity of the city and make people jubilant'. In general grain and edible oils were provided for urban people in the same amounts as during the previous year's festival. However, for peasants it was a different story. People's communes and their units were told to prepare for the festival based on the principle of 'self-provision – self arrangement'. Their level of supplies anyway had to be lower than in the city districts. Only if they were missing some items, could the city use its discretion to provide them with what was needed, but nothing was promised.[182]

During the first half of 1960, much of city leaders' energy was directed at continuing the steel leap and establishing urban communes. There were now 47 urban communes in the city and Party Secretary Wan promised their consolidation.[183] On 17 January, the City Party Committee convened a meeting of finance and commerce, streets and cultural *xitong* leading cadres demanding promoting 'people's livelihood' through promoting and 'doing in a big way' the urban commune mess halls. All units and streets were told to promote them and the majority of residents were to join them, it was demanded. Urban communes also had to help to continue developing the socialist economizing movement. This had staple foods as its core, but also promoted economizing in coal, electricity, vegetables and cloth.[184] The City now organized People's Livelihood Committees to direct urban communes' mess halls, kindergartens, service stations and collective engagements.[185] Propaganda also began to frame food problems as a result of bad and wasteful foodstuff management by individuals, which mess halls could solve.[186]

Throughout 1960, propaganda claimed that dining halls were spreading like 'spring sprouts after rain' and households were discarding their old private kitchens for them. It was argued that in the mess halls 'people eat economically, eat full, and eat well' and the 'common mess halls are so good that small home kitchens just cannot compare to them'.[187] In reality, urban communes were generally speaking a failure and the policy was not popular at all among urban residents, many of whom tended to minimize their involvement in communal services such as the mess halls. At the end of the year, a report on the Nankai District Dongshengli Commune revealed how only 81 households of the 140 households in the commune had initially entered the common mess hall and in December only six remained as its members.[188] The mess halls did not bring more food to the table in the city. On the contrary, in March 1960 all processed food became rationed and the city also began to require grain coupons in restaurants.[189]

The Worsening Situation in the Countryside

While the city was trying out ideologically correct, but in practice useless, organizational arrangements to improve residents' food supplies, the countryside was sliding deeper and deeper into disaster. Epidemics of famine-related diseases in Tianjin countryside were now becoming rife. The spring 1959 intervention against oedema had been inefficient and collecting wild vegetables to substitute for grain in autumn 1959 did not solve the food problems either. New reports of oedema continued to come to city health authorities, who continued to respond to them with medical interventions. In October and December, medical teams were sent to cure oedema patients in Ninghe County. In December, Ji County dispatched 1,000 health personnel to villages to investigate the 'Four sicknesses', i.e. oedema, malnourished children, womb prolapse and amenorrhea (absence of a woman's menstrual period) there.[190] These were all famine-related diseases or problems. The teams set up special clinics to treat oedema patients and delivered extra food rations to communes, usually 1–2 *liang* of grain per day per head,[191] but, as before, such interventions were just a drop in the ocean.

A February 1960 report from Hebei Party Committee to the Party Centre revealed that the province had diagnosed 59,000 oedema cases in its counties.[192] Although the number was most likely a gross underestimation, the epidemic was real. The Tianjin countryside had its share of this. In March, a City Public Health Bureau report on the oedema situation in the countryside was dispiriting. According to it, 11 of the 12 counties in Tianjin had reported of total of 8,345 cases of oedema and even 149 'unnatural deaths' to the City Health Bureau.[193] The report was not frank on the scale of the epidemic, but it was more honest on the causes of oedema stating that it was due to lack of proper nutrition. Working-age people could be working 17 hours a day, which consumed a lot of calories, yet in some of the counties daily rations were only 2–4 *liang* of grain. In Ningjin County, adult people were able to absorb only about one-third of the protein they needed daily. Vegetables were also in short supply and the unhygienic ways in which 'wild vegetable' food was prepared caused diarrhoea. Oedema patients were also more prone to catching other diseases as well. The clinics set up by dispatched medical teams treated their patients for free with fruit sugar and extra rations. The report indicates that extra rations were also allocated to wider audiences in communes, which reportedly improved the situation there. The clinics do not seem to have been widely accessible though, as in Ninghe only 56 cases of oedema were treated this way,[194] which was a negligible number.

The largest of such interventions took place in March–April 1960 after Wan Xiaotang personally led an inspection team by bicycle to Wuqiao County. One person in the team described what he saw as 'wretched': people in the villages looked famished, staring blankly and moving without energy in the fields. In one village, the teams saw people crying over fresh graves, and the old, infirm, sick and handicapped could not stand the hunger and were dying in numbers. Farm animals went without fodder limping around with bulging bellies. The only food available was corn pulp and mashed sweet potato leaves and peels. These were served mixed together, making their eaters suffer from flatulence.[195]

In April, emergency intervention was decided upon by the City Party Committee to help Wuqiao, Ba and Baodi counties because of mass poisonings of the people from eating a poisonous weed called *cangerzi*. A temporary City Disease Prevention Office was set up under the City Health Bureau and 196 medical and cadre personnel were dispatched to counties with medicine and equipment. That the campaign was in fact about saving famine victims was shown in the type of 'medicine' served: extra noodles, rice, biscuits and milk powder. Reportedly, 'people welcomed this warmly' and many urban *danweis* supported the effort by providing transportation. Nankai University and the Disease Prevention Office together drew up materials to educate people on edible and inedible wild plants.[196] Officially, the poisoning case afflicted some 10,000 victims, 40 of whom died, although an internal report noted that only 90 per cent of victims recovered.[197]

This time other regions too such as Beijing, Tangshan, Shenyang and Shanxi sent help in the form of personnel and medicine, which shows how this disaster relief operation had central approval. Interestingly, in April 1960 Tianjin also participated in helping Shanxi to deal with similar poisoning cases by sending in fruit sugar and other supplies by airplane.[198] This suggests that 'objective' reasons, such as mass poisoning from wild vegetables that people were told to eat as a substitute for grain, could trigger even a centrally approved disaster relief campaign at the height of GLF mobilization. However, as before, it is clear that such sporadic intervention provided only short-term relief. The widespread oedema epidemic and 'unnatural deaths' were not solved by them and they did not even end poisonings. Hebei province had 802 poisoning cases by late September 1960.[199] Also in the worst hit Wuqiao County the problems continued. In June and September, more cases of mass food poisoning were reported with over 400 victims and in December eight communes in the county reported 336 people suffering poisoning from eating *cangerzi*.[200] The City Government office for handling the campaign was dissolved in May, which spoke volumes about the interventionist method. Basically the same office had to be re-established later in the year.

While the interventions were going on, the city actually tightened its grip on food production and distribution in the countryside. Already in January, the City Government issued an order that all vegetables grown in the city districts (apart from some vegetables from people's own plots) were to be sold to the state. In counties under Tianjin jurisdiction, all onions, potatoes, *baicai*, radishes and garlic had to be sold to the state as well.[201] On 1 March 1960, the City Party Committee held a telephone meeting on agricultural work conveying the provincial Party Committee's instructions that all grain that had been produced by production brigades had to be handed over to common mess halls (after the state had deducted its own considerable share). As a sign of the still ongoing anti-right campaign, the City Party Committee also instructed that private plots given earlier to peasants in a haphazard manner had to be returned to brigades.[202] During the summer harvest, which had been once again affected by the spring drought, the Grain Bureau still employed the 'buy great, sell great' method for wheat, which meant that the state procured all wheat produced (apart from seeds) and redelivered it to villages based on purchase certificates. All potatoes were also to be sold to the city vegetable company.[203]

Typical for the crop cycle, the city also mobilized a planting campaign for spring vegetables and edible oils and a campaign against spring drought. Preparing for the spring planting season, the City Party Committee and Government convened a seven-day vegetable production meeting for leading cadres to decide on production targets in February.[204] In March 1960, the City Government once again issued a directive to the urban masses for a campaign to use idle land to plant oil plant seedlings.[205] This policy continued to be unsuccessful as the production of edible oils actually dwindled throughout the year.[206]

The province and city put now forward a hefty slogan: 'every person grows 1,000 *jin* of tubers and vegetables'. In the countryside 210,000 people were mobilized for rushed vegetable planting at the end of April – early May. Urban units such as factories and schools were urged to plant vegetables on scattered idle lands. According to official statistics, the whole city was using 62,000 *mu* for this, which was 19.2 per cent of all sowed acreage. Chengguan Commune in Ba County was praised for using grave moulds to plant pumpkins in. Other units used the sides of roads, ditches and fields.[207] There was also a typical campaign against spring drought in the spring of 1960. Three hundred thousand peasants were engaged in the campaign, which continued until July. However, this all happened under instructions that fighting the drought must not take manpower away from other production duties and that localities had to find ways to compensate for the manpower, for example by using women in jobs where men used to work.[208]

The campaign to economize in grain consumption also continued in the city throughout the winter and spring of 1960. Yet another campaign targeting rations, *hukous* and restricting *mangliu* ('blind' migration from the countryside into the city) started in April.[209] The City Government justified the rectification drive by noting how rations had on average actually increased in the city and had to be cut back to normal.[210] In contrast, the Tianjin peasants' monthly ration was by now cut to 21 *jin* a month,[211] of which at least 70 per cent consisted of sweet potatoes.[212] Such a level could not sustain agricultural workers with heavy menial jobs. It was not surprising that voting with their feet attracted many peasants who ventured into the city to find food and better employment. Indeed, an early 1960 party document indicated how the inflow of *mangliu* was increasing constant, blaming the influence of bad classes in the countryside for the phenomenon. Allegedly, members of the bad classes had made people fear hard work in the countryside and look for better jobs in the cities.[213]

Propaganda was now framing famine refugees as one of the main cause of problems in the city. The City Bulletin admitted in April that there were food queues in Tianjin, but these were allegedly caused by migrants. The 400,000 or so temporary people in Tianjin were negatively influencing the market situation in the city, it was complained.[214] In August, the tone grew even harsher when the

Bulletin claimed that 'a majority of the people who claim that there is not enough food are from this group'.[215] The message was that a better food situation was attracting these selfish drifters to the city. Urban grain rations were much better than in the countryside for sure, but the spin here was exactly in emphasizing this fact, not the fact that in many places in the countryside rations were not enough to survive. This made the refugees appear selfish and jealous, not people driven by basic survival needs as was often the case. Indeed, oedema and other famine-related diseases were common with the in-coming refugees. The Tianjin Shourong and Repatriation Station now had to establish a special hospital to treat its patrons in order to prevent contagious diseases from spreading – but it took until 1962 for all district-level *shourong* stations to have clinics.[216]

Urbanites also caused problems. A big headache for authorities was that too many people were getting food they were not entitled to because many work units engaged in assigning people with higher grades than their work actually entitled them to. The amount of people classified as doing semi-strenuous or strenuous work increased from 338,000 to 400,000 in the first half of 1960 making the Grain Bureau complain that the controls were too lax in many work units and that they needed to be rectified.[217] The City Bulletin continued to propagate different models for grassroots organizations to save grain through peer-inspections and pressure.[218]

In May the Vice-Director of the Finance, Grain and Merchandise Office, Hu Yunchang, noted how since March 1960 the city had conducted rectification of rations for 600,000 people downwards and for some 50,000 people upwards. The average grain ration was now 29.7 *jin*.[219] The quality of food supplies also deteriorated further during spring 1960. In March 1960, up to 90 per cent of rations were 'augmented'.[220] In August, the City People's Congress declared that 80 per cent of mess halls in the city were augmenting their food.[221] Vegetable supplies went down to 2–3 *liang* per day, but this improved to 6 *liang* when the spring crop arrived.[222]

The summer crop was a failure in many places in China and in the first half of 1960 Tianjin received only 65.75 per cent of planned grain. The Grain Bureau urged its representatives in the provinces to work hard to get their procurement tasks fulfilled, sending them telegrams to this end.[223] In early June 1960, at the low point of supply cycle, the city only had grain reserves for ten days. Retail shops sold grain only in 3–5 day portions. Residents were standing in long grain queues for hours on end, sometimes getting nothing. Complaints were common. The situation was mitigated temporarily in June when a central emergency directive ordered grain-producing provinces to ship in 'surplus' grain to big cities including Tianjin. A representative from Tianjin also took part in discussions that preceded this central government decision.[224]

For his part, Party Secretary Wan Xiaotang made sure that other localities carried out this and other similar central orders. He dispatched his leading cadres to supervise grain shipments. At different times, these delegates travelled to all 22 different provinces or areas that provided Tianjin with grain. Even if the delegates could witness how the situation was far worse in places where they procured the grain that Tianjin demanded, such as in Jinan and Hefei, they could still demand grain for Tianjin based on the central orders in their pockets.[225] Thanks to the central decision in June 1960 and the summer harvest, the flour ratio at urban rations went up to 60 per cent in July – although people in the suburbs had to settle for half of this. Rice was not supplied in July at all.[226]

When there was little else to offer, the number of propaganda slogans on saving foodstuffs multiplied in the autumn and summer of 1960. A column in the City Bulletin in April 1960 contained 12 such slogans at the same time: one had to be 'overcautious and indecisive' in using grain, had to understand that 'small streams grew into big rivers', had to 'allow for unforeseen circumstances' in saving grain, had to exercise 'careful calculation and strict budgeting', be 'economic in eating and frugal in using food', 'put aside reserves a bit more, eating food a bit less', 'be frugal in the days of aplenty', 'use grain in a planned manner', 'save grain', 'have hope', while nevertheless 'eating one's stomach full'.[227] If such slogans alone could have filled the empty noodle bowls, Tianjin residents would have had nothing to worry about.

The Emergency Phase: Late 1960 – Summer 1961

Starting in July–August 1960, disaster management can be said to have entered an emergency mobilization phase in Tianjin. Thus far, various campaigns that had been part of crisis management had been limited and had not interfered with production duties in a major way. More active disaster management became possible when the Party Centre began its final withdrawal from the Leap, but at the same time consecutive central government decisions and the worsening grain situation in the whole country led to a real prospect of famine also in urban areas. In propaganda the Party started, belatedly, to pay more attention to the 'livelihood' matters of citizens, which was a euphemism for disaster relief measures.

The 1960 July–August Party Central Committee summer meeting in Beidaihe is usually regarded as the beginning of the second and final withdrawal from the GLF. The meeting decided to curb basic construction and re-allocate manpower to agriculture. Already at the end of July, the Tianjin Party Committee convened its own meeting of all city party cadres to study the spirit of the Party Centre's decisions on this matter. Party Secretary Wan explained the new line of 'guarantee grain and steel production'.[228]

Already before these decisions were made, the central government had cut grain rations in big cities. In the second half of 1960 grain supplies from the major grain production bases in Heilongjiang, Jilin and Sichuan dried up altogether and the central government had to resort to central granaries to support deficit regions, which began to deplete national grain reserves quickly.[229] As a consequence, the central government decided to cut urban grain rations. On 12 July, the Tianjin Party Committee and Government followed central directives on this and issued an order to 'economize grain consumption, prepare to get over a lean year', which stipulated that everybody in the city had to cut their average monthly grain rations by 8–10 per cent.[230] People, who had been saving grain since early

1959, were now told to 'save grain with all their strength' starting from the second half of July.[231]

The order also noted that cadres should go down to mess halls to enforce the regulations and see that workers and staff members were 'eating well and their full' (*chibao, chihao*). Mess hall cooks' and management personnel's ideological education had to be strengthened. Food augmentation methods had to be further promoted, while 'amounts, variety and quality [of food] had to be ensured'. All food outlets and city and countryside mess halls were to be mobilized to use idle lands to plant vegetables for their own usage. Suburban mess halls were once again instructed to collect 'wild vegetables', and use more ersatz foodstuff, while avoiding the use of market vegetables, which were designated for city use.[232] It was further ordered that all grain saved in *danweis* had to be managed by grain bureaus and used only under the orders of the City Party Committee or City Government, and categorically not to be used without authorization.[233]

While the Central Committee Beidaihe meeting was still going on, city leaders also stepped up the ongoing supplementary food production campaign in late July, when the autumn planting season began. Now that manpower restrictions set by other campaigns were easing, the City Party Committee and Government held a joint meeting on 13 July to organize a mass campaign for autumn vegetable growing. A very rare green banner on the front page of the *Tianjin ribao* called upon everybody to 'Seize the day, plant autumn vegetables in a big and special way!' The article conveyed the earlier provincial meeting spirit on improving urban vegetable supplies, and explained the reason for the campaign by the already familiar reason of constantly growing demand in the cities, but also added by way of explanation a historically rare drought, which had influenced vegetable production earlier in the year. All localities now had to make vegetables their top priority, mobilize the masses and use all available means to fulfil or exceed the production targets for the year. People's communes, especially the ones in in suburbs, were told that guaranteeing city market supplies of vegetables was their 'glorious task'.[234]

According to the City Party Committee's order on the mobilization of the city for a 'great vegetable planting campaign', each work unit had to plant a fixed amount of vegetables per person.[235] They had to use all idle land for the purpose. This could be the sides of residential buildings, the sides of villages, ditches, raised platforms next to wells, etc.[236] A leading cadre work group of 70 persons was selected to lead the campaign and responsibilities were divided between districts and counties. All urban units were urged to establish special offices to handle the implementation of the policy.[237] The highest city leadership clearly regarded the campaign as important. In a speech to leading cadres in November, Party Secretary Wan described vegetable growing as the 'the way out' of difficulties in all areas suffering from food shortages.[238] This was a real mass campaign. By 1962, 3.5 million people in the city and its countryside had been mobilized to grow vegetables on unused land.[239]

The policy also achieved at least moderate success. The acreage used to grow vegetables grew from 170,000 *mu* to 230,000 *mu* in 1960, some of which came from the mixed planting of vegetables with grain, cotton or trees (i.e. in spaces in between them).[240] The Supplementary Food Bureau was able to purchase 7.15 million kg more vegetables in 1960 than in 1959,[241] but on average this result meant that urban residents got a paltry 0.5 g more vegetables per day. However, one needs to be careful in assessing these figures. What they indicate, is that average hectare yields were declining drastically in 1960. Therefore one can read the figures as indicating that the policy was able to offset the decline in the absolute production quantity of vegetables. Following this logic, if the city had used the same acreage for growing vegetables in 1959 and 1960, production would have fallen by about one-third in 1960. Vegetables made up about one-third of urbanites' food rations, so the result can be said to have made a difference for them. Further, work units would also have disincentives to report their own production truthfully to the Supplementary Food Bureau in order to keep as many vegetables for themselves as possible. The same was true of individuals, of course, so the actual production figures were probably higher. In November, the urban *danweis* were given the right to keep up to 7 *liang* of self-produced vegetables per staff member. The rest of the produce was to be handed over to the state.[242]

Nevertheless, considering the effort put into growing vegetables, the results in 1960 must be regarded as disappointing. This can be explained by a number of external reasons. The major problem was the general shortage of seeds, which City Government policy announcements dealt with repeatedly, but lacked any real solutions. Propaganda could not bring seeds to work units' bins. Work units were told to solve this problem by 'relying on the masses' and the 'spirit of *zili-gengsheng*', which was of no practical use to them. One practical way propagated was to plant seeds more sparsely. One commune was praised for having 'economized' planting so that it could plant 5,000 *mu* with seeds usually used for 4,000 *mu*. Peasants were also told to take care that 'every seed becomes a sibling'. Dibble seeding was recommended instead of scatter seeding. Even growing vegetables in pots was recommended.[243] The City Commerce Department also dispatched buyers to purchase vegetable seeds from other provinces, and communes were urged to try to exchange seeds with other units in good communist sprit.[244]

Just like augmenting food with water, planting seeds more sparsely was a typical Maoist economic gimmick. It increased the acreage under cultivation, but hardly the crops. It is difficult to estimate how much this counted for the total decline in hectare yields, as other factors such as lack of fertilizers and using excessively alkaline land ill-suited to vegetable growing could also be factored in. Further, not every *danwei* was eager to divert its manpower to such rustic activities, when they were still expected to fulfil their production targets.[245] Instead, they could make 'cooperation' agreements with rural production brigades exchanging their own produce for vegetables and other supplementary foodstuffs without the bother of having to grow it on their own. The City Government tried to prevent this practice in vain. This said, the vegetable production campaign helped to keep at least this part of food supplies constant when all the other supplies were going down. It is therefore clear that in the bigger picture, and especially starting from 1961, self-produced vegetables did contribute to the survival of urban residents in Tianjin.

Promoting ersatz food production in a big way was also part of the campaign. This had been stepped up already in June when the concept of 'synthetic', or combined use of raw materials became common in propaganda.[246] It was now propagated that tuber and grain poles, shafts, buds and tendriled vine could be turned into starch, sugar, alcohol and seasonings. Leftover chaff could be fermented and used as good fodder saving grain and developing pig farming, which was still official policy. Propaganda demanded that people's thinking had to be liberated and 'superstitions' (i.e. doubts) had to be erased concerning synthetic use. People should for example learn to think that 'making alcohol without grain and starch without food is completely possible'.[247]

At the beginning of July 1960, the City Bulletin called for a mass tide in the promotion and use of ersatz food production,[248] and in August 1960 all *danweis* in Tianjin were ordered to grow mushrooms and 'small ball' algae, which was one of the most propagated types or ersatz food.[249] Other types included leaf protein, chlorella and 'man-made meat'.[250] By the end of the year, schools would be reported as producing their own algae and 'man-made starch'.[251] The actual nutritional value of ersatz food has, however, been disputed. Much of it had hardly any calories, and some could even be harmful to consume.[252] At best, ersatz food, such as 'wild vegetables', could quash the feeling of hunger of urbanites and peasants alike, but when not prepared properly, it could make things even worse.

While the city was mobilizing its resources to promote vegetable and ersatz production, a central directive cut rations by a further 2 *jin* in September 1960.[253] Edible oil rations were cut so that workers and cadres received 5 *liang*, and street residents only 3 *liang* of oil per month.[254] Reacting to these decisions, the City Party Committee convened two all-city cadre meetings where Party Secretary Wan raised the slogan of 'saving grain to pass difficult times'.[255] Typically for the harvest cycle, reduced rations were accompanied by the worsening quality of rations as well. By November, the flour ratio sunk to 10 per cent.[256] These developments have been seen as the end of a relatively tolerable situation in urban Tianjin and the beginning of the descent into widespread hunger even there.[257]

Peasants in the suburbs were also now urged to save grain with outmost strength,[258] as if they had anything left. At the end of the year, Hebei province decided that peasants' ration would be 4 *liang* of grain – well below substance levels. Ersatz food was to substitute for the rest.[259] Tianjin leaders were acutely aware of the magnitude of problems in the countryside. Wan Xiaotang again personally visited Cang County in September 1960, witnessing how the food situation was 'extremely tense' and how experiments on planting rice on dry alkaline land had failed there. Even after the autumn harvest, the county had grain reserves left only until the end of the year. Wan visited many communes and households in the county all with same results: households did not have any grain left. He instructed the county to prepare well for '*duhuang*' ('tide-over lean year'), find anything that could be eaten and gather it. He is said to have been visibly very disturbed by what he witnessed.[260] In October, counties around Tianjin once again organized 'small autumn crop' campaigns to gather 'wild vegetables' and grass, and even *Tianjin ribao* advocated this.[261]

The city continued its policy of aid interventions in its counties. In August and September, Jiaohe, Jinghai, Cang and Hejian counties received relief aid due to flooding there. The City Party Committee also ordered the counties to mobilize a 'production for self-help' campaign. Victims were each given a 20 *yuan* subsidy and instructed to plant vegetables and wheat as well as prepare substitute foods to guarantee their food provisions for winter, while at the same time the areas were told to strive for a bumper harvest.[262] A *Tianjin ribao* article reporting on the city activities to help the areas even used the term 'refugees' (*zaimin*) for the only time during the GLF famine when it referred to those suffering from the officially recognized flood.[263] The word was never used for people fleeing the unrecognized famine, which once again underlines the political nature of the disaster.

July–August 1960 was also the time when natural disasters entered the famine propaganda explanation repertoire in Tianjin. Before this they were discussed as part of national news, as when mentioning big floods in the southern parts of the country in August 1959.[264] In connection with Tianjin itself, natural disasters were first mentioned in July 1960 when reporting on the anti-drought mobilization that had been going on since early 1960.[265] After this, natural disasters acquired a more prominent role in explaining the crisis, often combined with other already familiar explanations. In August 1960, the declaration of the second enlarged session of the City People's Congress claimed that at that time there was not only a great drought, but some places had suffered from water logging. It was assured that the struggle by the masses led by the Party against the drought had achieved a great victory. Nevertheless, because agriculture could not yet have recovered fully from these natural disasters, the province and the city had raised the slogan of economizing grains. The city received its grains from everywhere in the county, but as transportation had been relatively problematic due to long distances and the rainy season, spoilage had been high. These were now given as reasons for cutting grain rations in the city.[266]

Disaster propaganda could assume even more grotesque forms during this time. At the September meeting of the City People's Congress, the report by the City Party Committee Agriculture, Forest and Fishery Office used the drought to underline the great role of the Party and people's commune system in mitigating disasters by comparing it to the old society. It was claimed that this year there had not been rain for 300 days, making it a drought not seen in 100 years. Then the report added that 'Before the Liberation such a situation would have led to scenes of utter desolation and tragic sights of famine.' However, as the report argued, under the leadership of the Party, people were now able to fight the calamity. Moreover, the water works of past years had been very useful and therefore there had been a bumper harvest in summer crops. This said, the article continued, due to technological backwardness, agricultural production could not wholly satisfy the demand in the economy. In past two years, some locations had had shortages, and this year's natural calamities had caused definite troubles. The Party was taking the right measures in making grain production key in the economy, but in the meanwhile grain or cloth rations would not be increased this or next year. Instead, people should cultivate the spirit of 'hard work and thriftiness is glorious, luxury and waste is shameful'.[267]

Such propaganda was creating justification for further belt tightening in the city. Following central orders, city authorities began at this time to cut purchasing power in the city. In late September, the City Government issued an order to work units cutting their public expenditure by 50 per cent during September–December. This happened under the slogan of 'promoting expanding social saving'. The aim was to save grain, coal, petrol, cloth and paper.[268] A comprehensive saving campaign ensued in the city adding to the atmosphere of dearth. Most cars, including taxis, stopped running when sedans at lower than the city level were no longer given any gasoline. City-level offices and district Party Committee and Government leaders were still given 15 kg of gasoline per month for one sedan. Only cars in the use of the foreign travel service, emergency vehicles and sea transportation were given gasoline, but sparingly. All registration plates from cars that were not used were collected by the police.[269]

Reducing City Population

After the Party Central Committee had urgently ordered an increase in the agricultural workforce, Tianjin leaders made quick plans to reduce the population in urban areas and thus the number of mouths to feed in the city. The policy targeted both the refugees and rural residents who were living in Tianjin without a permanent *hukou*. The City Bulletin now announced that 'solving the workforce problem is the key to solving grain production problems'.[270] In the countryside, counties and communes were ordered to allocate their workforce away from non-agricultural works such as water works, schools or industries, but even theatre troupes and sports teams were disbanded and their members sent to support grain production.[271]

The reduction of urban population was outlined in September and October decisions. At a meeting of leading cadres in September, the Vice-Director of the People's Committee Finance, Grain and Merchandise Office Hu Yunchang noted how there were 210,000 excess people in the city. Sixty thousand of these were the relatives of temporary workers and those employed by urban households, 40,000 were *mangliu*, 40,000 were temporary visitors, and 70,000 were dispatched workers from other cities and localities. Speculative activities by such dispatched people were blamed for shortages in Tianjin generally, and they became a target of constant rectification campaigns during the crisis. All in all, it was argued that these outsiders increased grain consumption in the city and were out of agricultural production. A strict new policy that people 'can go out, but cannot enter'[272] was proposed by Hu, and it became the line that guided population policy in the city thereafter.

In October the City Party Committee decided on the repatriation of people from the city in six categories: 1) all *mangliu* without *hukou* or ration tickets, regardless of whether they were looking for a job or not; 2) peasants loitering in the city streets; 3) those who had come from villages and had a regular *hukou*, but lacked a regular job; 4) cadres dispatched from the countryside to Tianjin and their dependants regardless of their *hukou* status or work ability; 5) contract workers from the villages whose contracts had ran out and from companies where production was not urgent even if the contract was still valid; and 6) all people who had been recruited privately from the villages. All localities were ordered to receive the repatriated people 'cordially' and offer no resistance to the policy.[273]

The City Party Committee further argued that the number of *mangliu* was seriously affecting agricultural production and therefore capturing them and sending them home had to be stepped up. Although it was now admitted that some of the *mangliu* were fleeing excessively low rations and problems in their livelihood, this did not matter in the big picture. The city policy of receiving and handling the floating population was blamed as too generous with work assignments that paid compensation and offered too high rations. The new policy was that assigning temporary work to refugees had to cease. Instead, they had to be quickly received and handled. The railway station officials had to capture all *mangliu* they saw, it was ordered.[274] Following this, the City Government issued an order to 'strictly receive and promptly repatriate, make things clear in one day within the reception',[275] basically meaning repatriating all refugees ideally on the same day that they arrived in Tianjin.

To support reducing grain consumption and sending people back to the countryside, one more 'three checks' campaign to clarify and inspect *hukous* and to verify job classes and individual rations was carried out in the city in the autumn of 1960,[276] and the City Government once again ordered that companies be prohibited from hiring any more peasant labourers. Neither were they allowed to transfer personnel from their own operations from outside the city into the city, except by special permission, otherwise the police would refuse the *hukou* transfer, and the Grain Bureau would not supply grain tickets to the newly recruited personnel.[277] In October, the City Party Committee set the target of 120,000 ejections by December.[278]

Disciplining Cadres and *Danweis*

Party cadres who were involved in the food supply system now became a target of a discipline campaign that lasted for over a year. On 7 October 1960, the City Government secretariat held a meeting discussing the Party Centre's order on rectifying the responsible cadres' special supply system and curbing backdoor purchases of commodities. The City Government ordered that the use of the special supply system had to be strictly investigated. The proper functions of the system were to be preserved, but improper and excessive uses of it were to be curbed. Those parts of the system that were to continue existing had to cut down their supply by half. City leaders ordered that a mass campaign against using the backdoor had to be organized in the industrial and commerce *xitongs* and in other units that handled daily necessities. However, corruption, pilfering and lawlessly infringing against discipline had to be handled as separate issues from the 'general kind' of using the backdoor, and the capitalist elements' bad activities had to be separated from instances where cadres were being influenced by them. The basic principle in handling the cases was to be lenient on past errors, but strict on future ones.[279] When the Grain Bureau reorganized its operations in December, it was emphasized that only those who were politically reliable, honest in performing their duties, and working earnestly were to be selected to newly established positions.[280] Not only food, but also honesty, was in short supply.

Breaking the rules was systemic. In mid-October the City Party Committee convened a meeting for the cadres responsible for foodstuff and everyday commodity supplies to expose wrongdoers amongst them. It was declared that the City First Construction Bureau had defrauded 43.2 *wan jin* of grain; Dazhigu Granary had let 20 *wan jin* of grain go to rot; Heping District Grain Office had misallocated 'a lot of' grain. Five high cadres were punished for fraud, misallocation of grain and edible oils, corruption and plunder.[281] All party cadres and members were ordered to make self-assessments and submit themselves to public evaluation in order to cut their personal rations.[282]

The old problem of disciplining *danweis* also continued as ever. In November, the City Government reiterated its regulations on supplementary foodstuff price controls, warning *danweis* against going directly to villages to purchase their supplementary foodstuff there. It was especially noted that the practice of striking 'cooperation agreements' between production brigades and *danweis* had to end. It was complained that such practices made some producers not sell their produce to the state, but to work units at higher prices.[283] In this connection, the City Bulletin criticized by name the city Ethyl Alcohol Factory and Tobacco Paper Factory for striking cooperation agreements with production brigades in suburbs. In October, the Ethyl Alcohol Factory had purchased 6,000 *jin* of *baicai* at a price 100 per cent higher than state prices from a Jinghai County production brigade. When caught, the factory tried to claim the vegetables were from its own fields, and while still being investigated by the Price Management Authority, the factory went and bought another 30,000 *jin* of *baicai* from the same production brigade during the nights of 19 and 27 October. The Tobacco Paper Factory was even more heinous in its activities. It had also established cooperation with a Jinghai production team through a counterrevolutionary element sent down for *laogai* supervision there. The team got tobacco paper, filters and soy sauce from the factory and in return the factory used its car to fetch 20,000 *jin* of *baicai* from the team. This kind of 'departmentalism' and 'decentralism' was serious and had bad effects on the masses, the City Bulletin complained.[284]

The Nadir

After October, all counties under Tianjin jurisdiction began to report states of disaster to the City Government. Cases of oedema and people dying from starvation increased daily in the rural areas and oedema also spread to the city, soon becoming an epidemic there.[285] Finally, city leaders gave disaster management full priority. Party Secretary Wan addressed the crisis head-on in a speech to leading cadres on 9 November. He now saw it as the party committees' first priority to protect people from famine, prevent oedema and generate recovery. Echoing central instructions, he emphasized the slogan of managing people's livelihood, i.e. food, shelter, fuel and health issues. According to Wan, in villages about half the workforce should be dedicated to managing livelihood issues and only a little energy should be spent on water works and other tasks. In the city, all first party secretaries were to dedicate one-third to two-thirds of their working time to solving livelihood issues. Wan also stressed that livelihood questions were not 'general' or economic problems, but political problems, about the party–masses relationship.[286] Elsewhere, Wan is also said to have demanded there be no deaths due to famine in the city.[287]

Wan also saw the so-called *gua-cai-dai* policy, i.e. growing vegetables and promoting ersatz food production, as the way out of troubles in all areas, but complained how ersatz and vegetables were still produced in excessively small quantities because some cadres, units and communes were not implementing the policy well enough. In a minority of areas, especially disaster and poor areas, people were fleeing and 'irregular' deaths were occurring, while some areas were eating more than their rations. Some cadres lacked awareness of livelihood questions, especially in Cang County and Jinghai, Wan noted. He also had a very standard solution for this: on all levels cadres should go to the front line to mobilize and talk with people, while teaching the masses 'the truth' about the situation.[288]

Wan Xiaotang's belated call to concentrate city efforts on solving the crisis was given special urgency when the City Party Committee convened a leading cadre meeting to convey the central report from Premier Zhou Enlai on 'Problem of Shandong and Henan Xinyang Cadres' Illegal and Undisciplined [actions]' on 25 November. These reports dealt with the famines in their respective localities, where it had reached genocidal proportions. At the meeting, Party Secretary Wan named Cang, Jinghai, Ba, Renqiu and Wuqiao counties as the focus locations for rectification and the city dispatched cadre teams to investigate them. He now declared that 'life was a priority', even at the expense of class struggle,[289] which is said to have earned him the title of a 'counterrevolutionary revisionist' during the Cultural Revolution.[290]

In late November and December, counties and city districts held their own meetings to discuss livelihood issues in them.[291] In Hangu District, such a meeting was called the 'Eating Conference'.[292] On 10 December, the City Party Committee convened a general meeting to discuss the livelihood campaign. Now 'eating' (*chi*) was made the key to all rural work. Livelihood questions had to be a priority in every aspect of cadres' work, it was instructed.[293]

To manage the campaign, the City Party Committee established the Peasant People's Livelihood Office in November, while the Urban People's Livelihood Committees had been in existence since the beginning of the year. Their task was to organize residents' nutrition, rest and health, and to strengthen the management of those who were sick due to lack of nutrition.[294] The City also re-established the Disease Prevention Office at the Public Health Department, which was put in charge of improving people's health against oedema and malnutrition, which were now officially acknowledged as existing at least in internal publications, if not publicly.[295]

Following this, a city-wide public health campaign was launched in late November. According to the campaign instructions, health workers were to survey and cure the cases of oedema, typhoid fever, uterus prolapse and chronic stomach and intestine diseases as well as prevent food poisonings. In the villages, the campaign was also to deal with mess halls and their food hygiene and the nutrition levels of various foodstuffs. Peasants were to be educated to distinguish poisonous 'wild vegetables' from safe ones. In areas where cases of diseases occurred, simple bed places were to be established for patients. All levels from city to production brigades were to establish Disease Prevention and Health Protection Head Quarters or small groups headed by a secretary-level cadre to direct the campaign. An Urban Health *xitong* was to send personnel down to counties to support health personnel during the campaign there.[296] As a consequence, over 1,000 health personnel in 14 large teams were once again dispatched to the countryside to treat patients with famine-related diseases.[297] But the Party was not acting on any, belated, humanitarian impulse. As the *Tianjin ribao* noted, the health campaign was needed in the countryside to 'guarantee a firm basis for next year's bumper harvest'.[298]

The City vs Oedema

Fighting against oedema now became the main front in the city's emergency mobilization against famine. As described above, in late November Party Secretary Wan ordered *danweis* to screen for oedema patients and establish special clinics for them. The numbers of treated patients were to be reported to the City Party Committee first weekly, then daily, according to the three-stage oedema severity scale. The figures kept on rising rapidly. At the end of the year, 700,000 city residents had been diagnosed with oedema.[299] Even Wan Xiaotang could personally attest to the situation in central administration where he personally tested many cadres for oedema by pressing their skin. People whose skin did not bounce back after it was pressed were suffering from the disease.[300] A late December instruction from the City Livelihood Office noted that, when leading cadres found oedema cases in any units under them, the party unit in these units had to investigate the cases and then decide who was sick. People were no longer required to go to see a doctor for diagnosis. Street residents still needed a medical check-up by medical authorities to be confirmed sick, but such check-ups had to be carried out 'nimbly' so as not to hinder aid for patients, it was instructed.[301]

The policy of making party committees responsible for screening the sick was probably partly warranted by the fact the public health care system would have choked if it were to check all oedema cases in the city, but the decision had also political reasons, which were probably more pressing. This was revealed in a report from the Construction Bureau written in late January 1961. The report complained that some of the medical personnel in the work units under the Construction Bureau lacked proper political outlook and training to judge who was really sick. For example, in a brick factory a doctor had diagnosed 580 workers out of the total of 695 with oedema. Admitting them all to a clinic would have stopped production. Therefore, the factory party committee had re-examined the cases and found out that there were really only 39 cases of the disease.[302] Typically for the Maoist state, technical expertise could be replaced with political expertise. From the Party's point of view, cadres made better doctors because they knew what the diagnostics should result in. The political nature of the screening process for oedema patients suggests heavily that the real numbers of oedema cases in the city were probably higher than 700,000.[303]

Nevertheless, following the policy laid down in November, all *danweis* established clinics for their oedema patients. The City Livelihood Office also decided upon extra nutrition standards for oedema patients. On paper, these supplement rations included 2 *jin* of beans, 1 *jin* of brown sugar, 1 *jin* of jujubes, and 3 *jin* of fish.[304] A lot of people needed extra rations. In December urban people were getting 10 per cent of their rations in flour and 15 per cent in rice, and the rest was given on an 'eat whatever there is' basis.[305] Average supplementary food rations for a month included only 3 *liang* of meat, 2 *liang* of fruit sugar and 3 *jin* of fish as well as some biscuits and fruits.[306] Clinics' extra rations, if the standards were followed, doubled these amounts in practice, but there were reports that not all clinics had such resources or could not get them because of shortages. Instead, they served their patients ersatz food.[307] Reports on how to manufacture various ersatz foods and on experiments in how ersatz food, such as the green ball algae, could be used to cure oedema patients were circulated internally.[308] Treatment also included 'western medicine', basically fruit sugar. However, when medicine was in short supply, only the seriously sick were treated with it, and the rest had to resort to 'local methods' (*tufa*), such as jujube soup, pepper soup, rubbing hot wine on skin and drinking hot water soup to make one's feet feel hot.[309] All these were actually traditional methods of treating flu.

It was further instructed that oedema patients could be allowed to stay home, but clinics were preferred because officially they made medical check-ups easier. In reality the reason was more calculative. Cadres enforcing the policy soon discovered that home patients could give their extra rations to their hungry children or other family members. This was out of question, since the idea of extra rations was to enable patients to resume their production duties. Clinics were indeed discovered as more effective in curing the patients than homes.[310] It was obviously easier to ensure that patients consumed their rations themselves in the clinics than at home. As so often was the case, economic, not humanitarian, motives were behind communist relief aid.

The basic thing patients were to do in the clinics was to rest. Indeed, it was soon discovered that resting was the best way to cure oedema symptoms. For example, the City Construction Bureau's clinics instructed patients to lie in bed for four hours a day 'not smoking or talking' and to 'let their brains rest'.[311] The clinic room temperature was to be kept between 15 and 20 °C. Patients were to engage in some light exercise, given nine hours sleep per night, washed 1–2 times a week, and a medical check-up was to be performed at least once a month.[312] It is anyone's guess how well the work units could follow such standards during the mid-winter in a city feverishly trying to save fuel. According to Yang, in the countryside clinics were usually short of medical equipment, crammed and had no food, medicine or medical personnel.[313] In a city like Tianjin the situation was clearly somewhat better, but letting people rest, not high-level medical care in the clinics, was the key to getting better even there.

Tianjin schools also took action against oedema. The winter vacation time was brought forward and extended to give students and teachers a possibility to rest. Schools divided oedema patients into three types: type one patients, i.e. the most severe cases, were forced to live in school clinics; type two were to eat and rest at school during the day, but allowed to go home in the evening; type three were unrestricted. About 30–50 per cent of teachers belonged to types one or two, but most students were classified as unrestricted, and only about 4 per cent of them were diagnosed with oedema before the vacation, which was a very low figure compared to the working-age population. Reportedly, the vacation cured three-quarters of them, although some of these cases later relapsed.[314] In comparison, 38.2 per cent of students in the three major universities (Hebei, Tianjin and Nankai) in Tianjin were reported to be suffering from oedema at the end of January 1961.[315]

The policy of forcing people into clinics had a number of problems. Reports complained about patients who were using their resting time to loiter around the city and engage in black market activities. Some patients demanded more food, while some did not want to leave after being cured.[316] In schools, many female teachers fought against the policy of living at school, because they could not take care of their children if they had to stay at school. Therefore only a few female teachers actually did as the City Government expected them to.[317]

In response to the problems and the general crisis, clinics were urged to carry out propaganda work on their patients, telling them that curing oedema and keeping healthy was important for the Party and the revolution. Patients were told to overcome fear of death. Schools also told family members of oedema patients how to treat them.[318] The Construction Bureau reported how the political education of the patients had been stepped up because some of the diagnosed comrades had been ‘unhappy with the situation’, while others had ‘confused’ ideas about the reason for the epidemic and even feared dying. As part of the propaganda effort, clinics also offered entertainment in the form of card games, walks outside and movies to lift the spirits of their patients.[319]

For the rest of the population not given diagnosis that entitled them to treatment, the city followed the provincial policy of ‘striking a balance between work and rest’, which started already in November 1960.[320] *Danweis* were told to implement a ‘uniting work with rest’ policy, whereby weaker persons were allowed to work for half a day and rest for half a day or even for the whole day.[321] In many cases, this just acknowledged the existing reality. The policy meant, in essence, that the city was trying to control how many calories peoples’ bodies burned since it could not increase their calorie intake. Letting people rest made sense in the context of the famine and was actually important for survival, since many famine deaths during the GLF were due to making people with low or no nutrition work too much, especially in the countryside.

But many people got sick anyhow, so who were they? The available reports are somewhat counterintuitive. During January 1961, the Construction Bureau reported its oedema situation daily to the City Livelihood Office. The numbers peaked on 23 January at 17,861 diagnosed cases out of 90,958 employees, or 18.24 per cent, of them, after which they started to go down. Different offices and units of the bureau reported very different figures. The highest proportion was nearly 90 per cent of the staff diagnosed with oedema, while the lowest was about 1 per cent. More interestingly, however, was that, according to reports, cadres and administrative staff were more likely to get sick than other personnel groups, including common construction workers. A report in April 1961 claimed that in January–March 1961, 30 per cent of cadres, and only 6 per cent of heavy workers were diagnosed with oedema, and the higher one went in the administrative hierarchy, the higher the rates of oedema that were reported.[322]

Reports gave several reasons for cadres' elevated likelihood to contract oedema. Cadres were said to live more often at home, where uniting work with rest and the planned use of rations was more difficult. Their travel distances between work and home could be longer than for other workers and home affairs required energy after work. Cadres could have large rations, such as 46 *jin*, but with large families the family would eat all the food before the end of the month. Among cadres, there were cases of families that had no food from 5–16 days a month. Therefore, all of their family members could be oedemic. In contrast, many contract workers lived next to work sites and could rest in barracks, while having all their rations to themselves. Also the occurrence of chronic diseases and the way mess halls and oedema clinics were managed in units mattered, it was argued.[323] While the given reasons may have been valid as such, the result is nevertheless counterintuitive. Cadres had better rations and did not engage in physical work. One cannot but think that the highly politicized nature of the diagnosis process probably played a role in the results too. It was, essentially, cadres, not medical professionals, diagnosing other cadres. Whatever the reason was, the result was that urban relief aid went disproportionally to cadres.

A report from the grassroots level suggests that it was indeed not the cadres who suffered most from food shortages. A survey of Nankai District Dongshengli urban commune revealed that at the end of 1960, residents' average grain ration was only 24.02 *jin* a month, that only 19 per cent of households had 'relatively abundant' food supplies and that 59 per cent had food until the 27–28th days of each month. Quite alarmingly, 26 per cent of commune members lacked food for a week or longer every month and 5 per cent of households were 'seriously' short of food. The most common period of having no food in a month was eight days. People had options of buying food, but queuing was a big problem, because one had to stand in a queue at midnight to get something. According to the survey, the best supplied were those households where all adults had jobs in industrial *danweis*.[324]

If the survey result was somehow representative of the whole city and its population of over 3 million, then about 150,000 people were 'seriously' short of food even by official standards. The 700,000 people diagnosed with oedema correspond roughly to the 26 per cent of population lacking food at least for a week. If the situation had continued for long, the prospect of mass starvation in the city would have been real. People who did not have energy to stand in a queue for food for age, health or other reasons, or did not have money to do so to begin with, were on the verge of starvation. For some of them, the clinics' help, if it was forthcoming, was probably life saving. But it is very unlikely that some of the urban people did not die from the combined effects of hunger and diseases in the city. Some people were also ready to kill for food. In May 1961, a young man named Han Wen, a *laojiao* ('labour camp') escapee and drifter from Liaoning had returned to Tianjin after being captured there earlier and repatriated. Han tried to strangle an old man with a rope in a *hutong* for a yellow croaker fish, which the old man had just bought from a market. The robbery failed, and Han was caught trying to escape the scene. He was given a death sentence.[325]

Typical of the unequal nature of Maoist society, some groups and individuals got better treatment than others. Party Secretary Wan saw to it personally that the police and *Tianjin ribao* propaganda workers were given extra food subsidies of 2 *liang* of grain. This was justified by the increased need to keep public order in the crisis and the role of propaganda work in soothing public sentiments. Cadres at and above office level were also given an extra bean and white sugar subsidy in order to keep them healthy while administrating the complex crisis they had created themselves.[326]

Extra subsidies could also be given to individuals. In late December 1960, the City Government Foreign Affairs Office ordered that an American English teacher Liu Di at Nankai University be given extra fish liver oil and more flour in her rations than others. However, it was instructed that people should not be told that this was because she was an American, but because she was an intellectual. In memoirs about Grace Liu, or Liu Di, her family members tell how they too suffered from hunger and symptoms of oedema just like many others in the city, but Grace herself belittled the 'tough times' during the GLF.[327]

In the city, the oedema epidemic was brought under control gradually during the spring of 1961. The City Party Committee ordered that the disease be 'basically' cured in the city by Chinese New Year in February and eradicated by April. Reports indicate a gradually declining number of oedema diagnoses during this time. In April, only a few new cases were reported, and almost all old cases were reported to have been cured.[328] Oedema, however, did not disappear from the city totally. Still in August, Wan Xiaotang described the oedema situation in the city as 'serious',[329] but he may have been talking about the city's remaining countryside as well. As late as in December, a City Government Livelihood Office meeting decided to establish a united reporting system for 'infirm and sick' adults and children. The decision was about oedema, because it came with detailed descriptions and pre-printed slips for diagnosing and reporting oedema cases.[330]

While it was mobilizing its resources against oedema, the city also continued to make sure that others did not get them and therefore shut its doors to outsiders even more tightly. In December 1960, the city ordered an almost total halt to inviting 'common domestic visitors' and giving them presents. Only diplomats, foreign guests and *huaqiao* ('the overseas Chinese') could still be invited into the city. Concerning visitors from above, it was strictly forbidden to organize banquets, or other 'special entertainment' for guests. They would be given food according to normal regulations. Banquets for different reasons, like meetings and similar extensive waste, had to be curbed. The heinous practice of 'going through the backdoor' in disguised form and eating and drinking big by inviting guests and giving them presents especially had to be curbed. The number of meetings also had to be reduced to a minimum in the city.[331]

In January 1961, the ongoing repatriation campaign for non-*hukou* residents was extended to cadres and dependants of military personnel stationed in the city without approval. Only those who had special approval by city-level bureaus, the province, or the central government would be given a city *hukou*. All district police stations were told to strengthen their inspection work on *hukous*. Cases where people had not applied for a *hukou* or had taken someone else's place by assuming his or her identity had to be investigated and handled properly. The dead, arrested and those sent to *laogai* had to be promptly removed from residential records. Newborns were not to be registered twice. It was further ordered that the control of temporary residential permits for visitors had to be strengthened and no stays longer than two weeks should be allowed.[332]

By the beginning of 1961, Tianjin had thus become virtually a closed city, but refugees kept pouring in and with them followed famine deaths. During the first ten days of 1961, 3,000 refugees from the Northeast, Shandong and Henan were detained at Tianjin Railway Station alone. Most of these refugees had oedema and 14 of them died after disembarking from the train.[333]

The Countryside Bites the Dust

In the countryside the situation was even worse. The medical teams dispatched to counties as part of the health campaign reported back distressing situations and also other alarming reports reached the City Party Committee. Tianjin got its own Xinyang in December, when a report by the Health Bureau team dispatched to Cang County reported 'serious' mortality in the county. Death rates had doubled there since the same time a year ago. In 1959, from July to December, 296 mortalities had occurred in the county, while in 1960 the same figure was 577, and this did not include children. Almost 300 people had therefore died 'in excess' due to malnutrition, as the report attested to.[334]

According to the report, the death rate in the county was highest in elderly people who lacked proper care, were malnourished and usually were suffering from two or three different diseases. Another group where mortality was high were oedema patients, which probably referred to working-age malnourished peasants. They often died suddenly, some of them 'stunned' after suffering from oedema for a time. An injection of fruit sugar could revive such stunned people for a while, but they could still die a couple of hours afterwards. If oedema patients had to do strenuous work, or were exposed to cold temperatures, or just did not eat for a while, their blood sugar would drop and they would die suddenly.

Only rest and nutrition could cure them. The report called the situation a result of activities by 'bureaucrats and bad persons' in the county. The City Party Committee circulated the report within its organization and commanded all cadres to take swift action to prevent such incidents from happening again.[335]

But this was too late for many. Another alarming report was received from Wuqiao county in December, when an anonymous writer from the county Public Security Office informed the City Party Committee about the unnatural deaths in the county. The Provincial and City Party Committees sent a joint team to investigate the case, and their report in early January indicated that 315 irregular deaths had occurred in the county in 1960. The final Wuqiao County report on rectification in late March indicated that about 10 per cent of the cadres in the county had made illegal use of disciplinary measures on people. The report listed 40 different methods of punishments/coercive measures used by the cadres, ranging from hitting to hanging, making people stand in snow, and making them drink urine. Methods also included cold and hunger, starvation and withholding work points (which virtually had the same effect as starvation). All together 15,036 people had been denied food in different ways and 183 had been 'caused to die' this way or another in the county, according to the report.[336]

Wan Xiaotang visited Wuqiao County personally in February to see the peasants' disaster situation there. He also toured all the eight poor southern counties in 1961. The four northern counties were regarded as better off and Wan paid less attention to them. The situation was worst in Jinghai, Cang, Wuqiao and Huanghua counties.[337] According to Brown, a January 1961 report to the City Party Committee noted how 210,286 people had suffered from oedema in the city's counties and over 2,000 had died of it.[338] The number was clearly an understatement, since it was not possible that the countryside with more people, hard menial work and smaller rations had lower oedema numbers than the city, although the number may have included only the most severe cases.

Foreign Grain Saves the City

1961 began in a tense atmosphere. Peasants looted granaries in the western suburbs at least nine times. In the southern suburbs, over 200 peasants raided a granary in December 1960, killing a guard. In the city, food shortages drove people to desperate measures and 50 cases of looting grain and supplementary foodstuff were reported in January alone.[339] While people were literally struggling for food, the Party Centre was finally dismantling the Leap. In January the Central Committee decision on making the 'eight characters'[340] the basis of economic work was also circulated in the Tianjin Party. The decision marked a further withdrawal from the Leap, which the city duly followed.[341] The central government also made other important decisions that helped more directly to handle the acute food crisis in big cities.

During late 1960 and early 1961, city leaders continued wrestling with grain shipments. The situation was so bad that in December the city only had grain reserves left for three days. Party Secretary Wan demanded daily reports of grain situation in the city and relayed these to the central government in Beijing, making daily telephone calls to either Zhou Enlai or Minister of Finance Li Xiannian.[342] Wan was afraid that if anything went wrong with grain consignments, there would be chaos in the city. He even made the City Party Committee write an urgent report to the Party Centre and sent Vice-Mayor Song Jingyi to Beijing to ask for more grain from central government. This lobbying action is said have solved the immediate problem in the city.[343] Some improvement in the diet already came from the central decision to substitute urban residents' rations with 0.5 *jin* of soya beans at the beginning of 1961.

Then, although denying that any famine existed, China began to import wheat from the West. Tianjin was one of the biggest beneficiaries of this policy change. The city had begun to prepare for the imports in early January and the first ship delivering grain arrived at Tangu Harbour on 2 February 1961 with a 11.62 million kg cargo designated for Beijing and Tianjin. The importance of the event was revealed by how the ship was received by a delegation led by Song Jingyi personally. In 1961, 1.234 million tonnes were imported through/to Tianjin, and by the end of 1965 this figure had risen to 5.621 million tonnes, of which 2.305 million tonnes came to Tianjin. In 1961, Tianjin received some 41 per cent of all grain shipments through Tangu, which was about 460 million kg of grain, or about half of what the city needed in normal times. Hebei province received about one-fifth of the imported grain, which was partly used for rural relief aid.[344] However, Tianjin had to lobby for this grain, as it was not delivered automatically to the city.

Although it was officially termed as foreign trade, the grain was in effect foreign relief aid. As a result of this intervention, city residents' rations were increased from the official average of 27.27 *jin* per month to 28.2 *jin* in early March 1961.[345] In April, the City Supplementary Food Bureau decided to distribute extra rations containing sugar, crab liver, bone flour and some meat. Doctors and nurses were also given an extra 0.5 *jin* of fish.[346]

More vegetables were now also becoming available due to the season, the vegetable-growing campaign, and a more relaxed policy on markets. In late spring 1961, city leaders decided on a trial of rural markets to improve supplementary food supplies. Produce not seen for years like eggs and some vegetables appeared in city markets once again.[347] But prices were high. Another change in central policies that had a connection to food supplies was the promotion of the sales of 'luxury' products, which began in January 1961 based on central directives. These included the sale of highly priced cakes and candies. One *jin* of such cakes cost 5–12 *yuan*. Highly priced restaurants also opened in Tianjin and, from March, 63 types of highly priced alcoholic beverages went on sale. For example, *maotai* cost 16 *yuan* a bottle.[348] One can question the wisdom of turning grain into alcohol under severe famine, but the measures were primarily aimed at curbing inflation by removing cash from circulation.

In the countryside, the situation remained bad throughout 1961, but the City Government continued its efforts to manage the disaster there. In mid-January 1961, the Party Committee decided to send 300 million kg of relief aid consisting of grain, ersatz food, coal and daily necessities to its rural areas.[349] The city also continued its efforts to curb the oedema epidemic in the villages. In mid-February 1961, the City Party Committee adopted the slogan 'stop two, change one' i.e. stop healthy people getting sick and sick people getting worse, while achieving a turn for the better in villages in one month, recovery in two, and resuming productive activities in three.[350]

Under this policy, the public health campaign continued in the countryside with a springtime drive. The instructions for the campaign had the same emphasis as before of curing people from famine-related illness, providing nutrition for them and letting them have enough rest. Measles was now seen as the main epidemic threat to prevent, but also diphtheria and even smallpox caused worry. Promoting personal and environmental hygiene was stressed. People were to be taught that 'to take care of hygiene is glorious, not to is shameful'.[351] Over 500 health-care workers were dispatched to villages in the second wave of the health campaign.[352] They were given a 0.3 *jin* grain bonus a day.[353]

The slogans probably sounded laughable in the worst-hit counties such as Wuqiao and Cang, which Wan Xiaotang visited personally in January.[354] Implementing the policies proved slow and the oedema epidemic remained acute throughout 1961 in the countryside. Party Secretary Wan continued to visit villages personally, witnessing how the situation remained bad in many places. In his visits, Wan even inspected people's excrement. Children's faeces were green showing that they were eating a lot of 'wild vegetables' which were difficult to digest. Wan also inquired about the situation of grain reserves, people's nutrition levels, starvation figures, what famine and disease prevention measures had been taken and what plans counties had to survive the 'temporary shortages'. Wan ate the same food that people were getting in the disaster areas such as 'wild vegetable' bread and corn starch and brought this bread back to the City Party Committee and let its Research Office people taste it, in order to 'educate intellectuals and cadres on people's difficulties'. Wan demanded that counties made sure that people get at least 0.5 *jin* of grain and disaster areas had to guarantee at least 200 g of grain a day to people. If a county did not have enough grain, it could seek it from sister counties and, if it was not available there, then it had to apply for it from the city and/or province, it was instructed.[355]

These amounts were meagre and would not sustain adults who were working full-time, so letting people rest was also part of the cure in the countryside. The Hebei Livelihood Office circulated instructions on how the sick should be given porridge, bean curd, meat soup, etc., foodstuff rich in nutrition and their rest times should be guaranteed.[356] But it is difficult to see where such foodstuff would come from. Local cadres were also warned that if oedemic people were made to resume their work too early, this could lead to their death. For example, in February, a Hebei Livelihood Office report on Wen County warned others how two county peasants had resumed work right after their recovery from oedema, going to cut wood in the forest and to a work in a colliery, but died on the road to work and on the hillside.[357]

Letting peasants rest was feasible in the slack season of January–February, but then the crop cycle required them to resume working once more. The tragic irony was that the peasants who suffered most from the famine also had to work most to stop the disaster and working under too little rations made them vulnerable to death by hunger. When the spring planting season began, disease conditions once again got worse in the countryside. In order to curb the epidemic, Hebei province decided to provide peasants suffering from medium or severe oedema (with doctoral or local clinics certificate) 0.5–1 *jin* of white sugar subsidy per month.[358] That this token action did not stop famine deaths was clear from the outcome. In the countryside, the cases of oedema and related diseases once again began to grow in numbers as did 'unnatural' deaths. This was all abetted by the enforcement of oversized grain procurement goals in the province that also continued in summer of 1961 and made some counties, such as Cang County, record growth in death rates *after* the autumn harvest there.[359] The Party Centre began to cut high grain procurement rates only in autumn 1961, which had a direct impact in ending the famine, but for Tianjin leaders rural famine was actually *passé* by this time because the majority of poor southern counties were removed from Tianjin jurisdiction between March and December. The Tianjin city leaders must have felt relieved.

Vegetables to the Rescue

The supplementary and ersatz food production campaigns continued and expanded throughout 1961, with one notable exception. In early January, work units were told that they were allowed to kill and eat pigs they had reared themselves.[360] This was an embarrassing end to the ill-conceived pig-rearing campaign that had been touted since autumn 1959. City annals do not tell how much this decision improved pork supplies in *danweis*, but otherwise Tianjin residents hardly ate pork, at least legally bought, in 1961 at all, as the *yearly* pork ration was only 350 g.[361]

The campaign to produce ersatz food was also in full swing. By spring 1961, ersatz food had become a commonplace dish in the city's mess halls and cafeterias. In March 1961, 1,352 *danweis* produced ersatz food and 80 per cent of canteens in the city served dishes made of or with ersatz food, providing 2–4 *liang* a day per person.[362] In the process of running down unprofitable industries, some factories were even turned into ersatz food factories.[363]

In late February and again in early March 1961, the City Party Committee and Government confirmed the continuation of the vegetable-growing campaign as a key policy to fend off hunger. In mid-February, the *Tianjin ribao* front page called for a campaign to plant vegetables that ripened quickly.[364] It also ran articles where units were described as having improved their vegetable supplies notably through striving to plant as much acreage as possible. The City Roll Steel Factory Number One was said to have been able to produce almost 110,000 *jin* of vegetables for its workers by cultivating 13 *mu* of land. The factory had used its workers with a peasant background to do the planting work and manufactured its own farm tools. Fertilizer had also been self-collected by assigned people. Seedlings had been grown in discarded pots in the factory.[365] However, another article reminded people that the planting campaign should not influence production.[366]

In late February and early March, the City Party Committee and Government held a two-week meeting organizing suburban vegetable production. As before, suburbs were told to 'take vegetables as the key' and take measures to increase vegetable production through an early spring production high tide. The meeting participants pledged on 'six consolidated' things: cultivated area, manpower, ploughing animals and tools, fertilizers, quantity and variety for vegetable growing.[367] The suburbs were told to dedicate their workforce, land, water and fertilizers primarily to this task. Production had to emphasise high-yield, quick-to-harvest and thick vegetables. Urban vegetable supply levels should reach 9 *liang* per capita per day and 95 per cent of the returned workforce was to be allocated to production teams that were to organize special teams for vegetable production. All districts and communes were told to quickly establish special leadership organizations to manage the campaign and to seize the day.[368]

Propaganda extolled how the suburbs were enthusiastically following the policy. One participating suburban cadre in the vegetable meeting was credited as saying: 'Making every effort to let city people eat fresh vegetables early on is our glorious task'.[369] Readers could read how some units had been expanding their vegetable fields and were planting their vegetables earlier than usual and how they were already pushing green leaves.[370] There was hope, at least according to propaganda, for urbanites, but what was not said was that even amidst the worst famine in the countryside *Tianjin ribao* was still calling peasants to serve the city first.

A major reorganization of rural administration took place in mid-March when the City Party Committee followed party central and provincial instructions to establish Tianjin Prefecture consisting of five counties, and Cangzhou Prefecture with nine counties from the counties that had been transferred under Tianjin jurisdiction in late 1958. Tianjin Prefecture was first jointly administered by the city and the province and consisted of Jinghai, Wuqiao, Ba County, Baodi, Ji County.[371] Together with the new Cangzhou Prefecture, these areas were the worst hit places in the Tianjin famine. Six months later, they were transferred directly under the province. The city was therefore relieved of a big administrative and relief workload. However, these areas continued to supply vegetables and grain to the city.

The Recovery Phase: Summer 1961–2

In Tianjin, the recovery from the famine can be said to have started in summer 1961. This means that food supplies were not declining anymore and in some sectors were improving. However, staple food supply problems persisted throughout 1961. This was the lowest grain production year in China and also in Hebei since the revolution.[372] Urbanites still had to wake up at 3–4 o'clock in the morning to queue to get food from shops.[373]

The general rhythm of recovery followed central decisions to withdraw from the Leap policies and reallocate manpower back to agriculture. Urban units' engagement in vegetable and grain production was systematized by the end of the year and connected closely to the reduction of the urban population. The City Party Committee and Government established Land Reclamation Headquarters (LRHQ) to oversee the implementation of the campaign. The LRHQ was to coordinate its efforts with the City Government Urban Population Reduction Office, which had now also been established.[374] However, 1961 also brought with it more unorthodox methods to promote vegetable growing by introducing material rewards.[375]

In June, rural mess halls were officially abolished by the Party Centre and the procurement prices of agricultural products were raised on a broad range. Tianjin leaders naturally also followed this cue closely. However, this was not without some embarrassment to them. The way the City had still been clinging to the original GLF policies was shown in how still in February the City Party Committee had rewarded 48 model mess halls at a big meeting.[376] A City history notes that abolishing mess halls was done explicitly to 'develop the capability of households to survive the difficult times', which was exactly the opposite to earlier propaganda on the superiority of mess halls.[377] In late July, urban communes were basically abolished.[378]

Once again in June, city leaders called for vigorous summer harvest and planting efforts. The *Tianjin ribao* called on all nearby suburban units to use all available cultivable land to increase grain production acreage. Only through expanding acreage could the city withstand natural calamities better, it was now argued.[379] In mid-June, the City Party Committee also discussed and circulated the new central decisions on saving grain and supporting agriculture by further reducing the urban population. Tianjin leaders followed the central line under the so-called 'double core' (*shuang ke*) policy of promoting grain production and reducing the urban population. For this purpose, the City Party Committee established a double-core small leading group.[380] The policy was summarized in a rather clumsy slogan: 'less grain [consumption in industry], less [urban] people, less draft to military, strengthen inspection work, improve work methods, strengthen political education, and arrange well the masses' living conditions'.[381] In early July, the City Party Committee outlined a plan to reduce the urban population by 300,000 people over the next three years.[382]

Connected to this, it was once again decided to check the urban population and food supplies, coined the 'double check'.[383] A specific office assigned to verify residents and their grain rations was set up under the City Government and 15,000 cadres were dispatched to units to carry out inspections. Completing their activities by November, work teams discovered 9,420 'black people' (i.e. those without a proper *hukou*) in the city, and 3,303 people whose *hukou* should have been rescinded, but for some reason had not been.[384]

Vegetable production policies became more systematic. In late June, the City Government issued Trial Instructions on Land Reclamation by urban work units. According to them, land that communes and production brigades were not able to use should be given to other units to cultivate, and communes and brigades should not try to obstruct this. Communes were promised some compensation for the land for three years. When growing grain, harvest from the reclaimed lands would be free from state procurement, and the produce was not to be computed into the units' own grain quotas. Plots over 100 *mu* were to be allocated by the LRHQ and its district sub-offices in a unified manner, and plots smaller than 100 *mu* were to be allocated freely by communes, brigades or units.[385] When the autumn planting season started in early July, the City Party Committee and Government once again issued instructions on increasing vegetable production at a big meeting that consisted of cadres down to production brigade secretaries.[386]

The new policies were slow to have a tangible effect on grain production and 1961 was still 'touch and go' for Tianjin grain supplies. Sometimes grain was still being sold out during the year in the city. Tianjin did not automatically receive its share of imported grain either, and had to keep lobbying for it. During the August 1961 Central Committee work meeting at Lushan, a leading cadre from the city's light industry sector named Li Zhongyuan attended the meeting as a Tianjin delegate in the Hebei group. The meeting also discussed the grain problem and decided to cut urban rations further. Tianjin, including its countryside, was proposed a cut of 70 million kg. Li knew that grain situation was very tense in Tianjin and the oedema epidemic was still serious. He reported the proposed cuts to Wan Xiaotang by phone, who after checking statistics informed that the city could not guarantee grain supplies for the winter even with present reserves. Li reported this to provincial leaders, but they did not mention the matter when it was their turn to report the situation the next day, because they felt that Tianjin had fewer problems than other regions. Following Wan's instructions, Li bypassed his provincial-level superiors and used the opportunity to talk about the city's supply problems directly with Zhou Enlai and Li Xiannian, pleading for extra beans. Zhou expressed concern and asked questions, 'solving the problem' for the time being.[387] But this was not a final solution and Tianjin leaders had to continue lobbying the Party Centre for grain.

In October, the instructions on land reclamation were made final and it was further ordered that the land-reclamation policy be carried out closely connected with reducing the urban population and downsizing *xitong* personnel through establishing suburban 'residential posts' on newly reclaimed land. The policy was now defined as to 'change non-peasant people into peasants and consumers into producers in order to not only lighten the pressure on the city, but also to solve partly the workforce needs of the land reclamation policy'.[388] The new residential posts were to be permanent settlements in the countryside for the downsized people. Part of the planned workforce was to be supplied by urban work units such as factories and schools.[389] The recommendation was that units should send about 3 per cent of their members permanently to work on land. Rotation was also possible. Normal workers should go first instead of technicians and the required experts should not be assigned to agricultural tasks at all. The quota was to cultivate 0.5 *mu* per person. Units had to arrange their own capital, farm tools, seeds and fertilizers for the task, although the LRHQs could assist them in getting them.[390]

In connection with the policy, suburban state-run farms were now ordered to reclaim land in a 'big way',[391] and help urban work units in reclaiming land and feeding their staff members before they could become self-sufficient. However, it was primarily the work units themselves which were responsible for supplying their sent-down staff members.[392] As an incentive, they were given an extra 0.3 *jin* of grain per day for the duration of their stay in the countryside.[393] The *Tianjin ribao* propagated the vegetable-growing policy and offered advice on what crops should be planted on what kind of land. The newspaper also criticized units that were not enthusiastic about using relatively inferior areas such as wet low-lying and saline lands for vegetable growing.[394]

In October, it was also decided that in 1962 urban units should be allowed to enjoy 100 per cent higher self-provision of vegetables from their own fields than state provisions (4–5 *liang* of vegetables a day in the winter of 1961–2), but they would not receive state rations during the months that they could support themselves with their own produce.[395] By the end of 1961, under the vegetable-growing campaign. over 2,000 *danweis* in the city became engaged in agricultural production using over 140,000 *mu* for producing grain and vegetables.[396] The vegetable-growing acreage rose to 305,000 *mu* in 1961 and vegetable supplies to the city increased by 27.3 per cent from 1960 (although the city had expected a 50 per cent increase).[397]

Furthermore, using material rewards for agricultural production became more systematic. In mid-August, the City Party Committee's agricultural work meeting decided to use material rewards to reach production targets.[398] What this meant was explained in the City Government decision in late November when it announced, based on the Party Centre's instructions, a detailed compensation scheme for grain, vegetable and edible oil producers. According to the regulations, grain production brigades got 10 *chi* of cotton cloth, one and a half cartons of cigarettes, and one pair of rubber boots for every 1,000 *jin* of grain they sold to the state. Soy beans, red beans and other kinds of beans, regardless of whether it was surplus produce or not, would get the same price. All grain the brigades sold above their procurement quotas were given an extra 50 *chi* of cotton, five cartons of cigarettes and three pairs of rubber boots. In edible oils, 100 *jin* of oil got five *chi* of cotton cloth in return as a bonus.[399]

As for vegetables, 1,000 *jin* of them got 100 *chi* of cloth, 15 cartons of cigarettes, and four pairs of rubber boots. For amounts that surpassed the contract regulations, each 1,000 *jin* would be given 150 *chi* cloth, 25 cartons of cigarettes and 12 pairs of rubber boots. Also, the surplus vegetables from commune member's private fields, which were sold voluntarily and had a sales certificate from the production brigade, would get the same price. Brigades were instructed to issue special 'reward purchase vouchers' by the City First Commercial Bureau and Grain Bureau to get their rewards.[400] Rearing pigs, poultry or other farm animals was rewarded with similar schemes. Fishermen were promised 1 *jin* extra grain subsidy for each 130 *jin* fish they sold to the state.[401] Indeed, the schemes meant that in 1961 the urban state began to practice virtual barter trade where grain and other foodstuff were exchanged for industrial products.

Compared to 1960, vegetable-growing policies were working better. However, as the policy rewarded production quantities, major problems with quality appeared, and large amounts of vegetables sold were of inferior quality or totally wasted.[402] Even with such problems, it is clear that this active vegetable production policy did influence positively the nutrition situation in Tianjin and thus contributed to the recovery. The irony of the policy was that the GLF was launched with the lofty goal of general industrialization. As it turned out, it rusticated Tianjin.

In the countryside the new policies and relief aid improved the situation somewhat. In 1961, peasants' average grain rations went from 21 *jin* to 27.5 *jin*, but there were still some 200,000 people in production brigades whose rations were only 21–24 *jin* a month.[403] Severely oedematic people were provided food subsidies by the Grain Bureau based on certificates from hospitals.[404] People were still eating wild plants and taking the risks that came with the practice. In Ji County, 681 people were poisoned and many people died because of food-related illnesses, including eating the canger plant.[405] In December 1961, the police had to dispatch officers and armed troops to bring the looting of foodstuffs in Baitangkou, Xianshuigu and Xiaozhan villages in southern suburbs under control.[406]

The Discipline Campaign Continues

Simultaneously with implementing the new agricultural policies, city authorities continued to wage an internal battle against what they regarded as illegal activities. By 1961, black markets had become more open in the city, running constantly around the clock, and moving from the more remote locations to the centre of the city, growing from scattered small trade to big business. In 1961, there were 13 black market spots in the city and all urban districts had their own black markets: at the Heping District's vegetable market, in the Hedong District's tunnels, in the Hexi District's People's Park, in the Nankai District's cooperative markets, and in a number of the Hebei District's public parks. The biggest black markets, such as the one in the People's Park, at any given moment had on average some 400– 1,000 customers and 100–200 black market traders.[407]

Articles on sale in black markets were vegetables, meat, birds, fish, eggs and fruit and some daily necessities. Prices could be 10 times higher than state prices on average, some products even more. 0.5 kg meat would cost 10–50 *yuan*, one chicken 15–22 *yuan*, an egg 1–1.5 *yuan*, and fish 3–6 *yuan* per 0.5 kg. Most traders were peasants from the suburbs, Hebei and Shandong, but urban residents such as workers, students, street residents and administrative workers did also participate. They were mostly speculators, with some self-producers involved too. Some sold products from their rations such as tobacco and sugar.[408] Even in the state markets, the prices were actually higher than list prices, but not as high as in the black market.[409]

The Commercial Bureau tried to curb such activities through occasionally coming down hard on them and engaging in political education of wrongdoers. Black market goods were confiscated, fines levied and the worst cases of speculators were sent to *laogai*. Regular market inspectors could criticize and punish violators on the spot and even hit them. In June 1961, the City Finance and Commerce Committee decided to recover market discipline by banning all collective markets that had been allowed for a while. Once again it was ordered that all vegetables, supplementary foodstuffs and agricultural by-products were to be sold to the state commercial and industrial bureaus. Industrial goods were to be sold only by state-run shops. Even foreign goods brought in by *huaqiao* were to be sold to the state, not freely on markets. People and units were also once again strictly forbidden to go directly to the countryside to buy their vegetables or 'choose tea leaves' from producers. It was complained that units unceasingly used this pretext to go and pilfer production brigades' vegetables and tools. The City Government decision claimed that such practices were 'making peasants very unhappy'.[410]

In September, 2,800 cadres were deployed to form inspection teams to enforce market regulations starting from the second week of the month when all black markets in the city were ordered to be closed down. This did not erase black markets, although they disappeared from publicity for a while. Even after the September rectification drive, underground black marketing continued.[411]

Also rectifying residents continued. Wrong work designations were as constant a problem as before. As a Grain Bureau notice complained in early 1962, in 1961 over 230,000 *jin* of grain was wasted monthly due to falsely reporting the type of work people were engaged in. Moreover, according to an investigation from January to June 1961, over 100,000 cases of different types of 'people-grain mistakes' had occurred, causing people to falsely receive 3.2 million *jin* of grain. Industry, commerce and fodder supplies had made similar mistakes worth 1.1 million *jin*. On top of this, the number of illegal activities of corruption, pilfering and reselling grain and grain tickets for profit constantly took place in the city. This was attributed to the departmentalism and individualism of many *danweis* and individuals as well as the activities of bad elements.[412] In the last quarter of 1961, the City Government decreed that all *danweis* should establish internal grain management systems to manage their grain work better and to be able to honestly check each month the number of personnel and their work types. They also had to establish exact recordkeeping of grain consumption and check it regularly. As usual, not all *danweis* responded eagerly to the call.[413]

Another rectification in 1961 targeted the party-state and the privileges its functionaries had taken during the crisis. In late September, the City Party Committee expanded the rectification scope to the whole city administration when it held a meeting specifically on cadres using the backdoor, i.e. using their connections to get food and other supplies. The meeting noted that 'the phenomenon of using backdoors is taking place constantly, in some case the situation is already serious. Especially one must note that some of the city and district-level leading cadres are engaged in this kind of activity.' The City Party Committee issued an order on 'strictly banning Party cadres from engaging in illegal black market activities' and ordered an investigation into the matter.[414] In December, a special leadership group for party rectification concerning using the backdoor was established led by Vice-Mayor Song Jingyi.[415]

On 3 November 1961, a mass trial was held at the People's Sport Stadium. A man called Wang Mude was tried for speculation and hoarding of state property and was immediately executed.[416] On 6 November, city authorities convened another meeting of 40,000 people to publicly condemn and punish 13 criminals charged with robbery, hoarding and reselling goods for profit, stealing and reselling rationing tickets, hooliganism and rape. Three of them were sentenced to death, and ten to prison sentences.[417] At the end of November, the City Party Committee issued a notice on punishing the Hebei Dajie Police Station Chief Li Fushan on his illegal and undisciplined conduct of corruption and waste, squandering state property, using the backdoor to acquire goods, using his office to blackmail people, falsely claiming state grain as his own, and dividing food up privately.[418]

These actions may have soothed public sentiments, but they did not solve the grain problem. Throughout the year, city grain reserves fluctuated between 28.96 million and 35 million kg. In November 1961, the city had once again grain reserves left for only three days' consumption. City leaders reported the situation to the Party Centre and their insistent lobbying finally paid off when the Party Centre decided that about 265 million kg or two-thirds of the national grain reserves in Tianjin be transferred to city use. Only this finally ensured that Tianjin would not sell out of grain again.[419] Workers, cadres and students were now given extra grain supplements to their rations,[420] as were the people suffering from sicknesses and children under seven. However, people in city suburbs were explicitly left out of these subsidies.[421]

From this time on, the acute food crisis can be said to have been over in the city, but although the grain situation finally stabilized, Tianjin was still a city in crisis. When the year ended, industrial production had declined in value by a staggering 50.4 per cent and agriculture by 8.3 per cent compared to 1960.[422] The general withdrawal from the Leap was evident for all: in industry, projects were scaled down drastically. Many companies suffering from lack of raw materials were shut down or merged with other companies. Over 200 construction projects were cancelled.[423]

Recovery Starts

Researchers still disagree on when the Great Leap Forward finally ended, as there is no single decision to the effect, but the so-called 7,000 cadre meeting held in Beijing between 11 January and 7 February 1962 has usually been seen as the meeting where the Leap was buried.[424] As usual, the decisions of the meeting in which Wan Xiaotang also participated were circulated within the Tianjin Party.[425] In early 1962, the Leap was therefore practically over also in Tianjin. The economic adjustment policy continued, as well as reducing the number of people in the city, stabilizing prices and 'restoring order'. Gone were the days of big-headed propaganda too. The City Propaganda Bureau held a meeting discussing problems in its work, such as boasting and superficiality.[426] However, this did not mean that the residents were now told the truth about the situation in the country.

Maoist agricultural policies continued to retreat. Now the 72 communes[427] in the Tianjin countryside were divided into 195 communes, cutting their size from an average of 7,000 households to 2,000, and teams were made the basic accounting unit. Many commune and brigade industries were halted, following the spirit of central policy on concentrating on grain production. By spring 1962, the Tianjin suburbs, like many communes around the county, had also begun to use household responsibility systems to improve productivity.[428]

In terms of grain supplies, 1962 was already better than 1961 in Tianjin, although many problems still remained. In 1962, Tianjin received 26.76 per cent more domestic grain than a year before and the trend was for the better.[429] The Livelihood Offices were abolished in May,[430] but the City Disease Prevention Office only in November 1962. The Public Health gazetteer notes that the general health of the population began to improve only in 1963 when the economy took a turn for the better.[431]

As usual, a typical mobilization for springtime vegetable growing began in late January. This time the propaganda emphasis was markedly on stressing not only expanding acreage and production, but also on the urban–rural relationship, which the GLF had strained. The *Tianjin ribao* ran articles, which described model units in suburbs doing their best to serve the city because of class friendship with the city people. A Hedong District suburban brigade was described as having voluntarily supplied the city with extra vegetables during the previous autumn's slack season because the commune's 'cadres and peasants had a deep class feeling towards the people in the city.'[432] Such propaganda usually described what was missing, not the reality. The people in the countryside had little reason to be thankful to Tianjin.

Policy on markets zigzagged. In June, the City Party Committee ordered that local peddler markets be allowed. Twenty-nine markets were opened, three of which sold urban products, and the rest agricultural ones. Each market had 100–400 sellers with vegetables, eggs and fruits as the main items. 60 per cent of the sellers were urbanites; 40 per cent peasants. Supply variety and amounts rose and prices stabilized in the city, and for some articles, like pork, the prices even went down. In a single week in August, markets would sell 15,000–20,000 kg of vegetables a day, 5,000 kg of seafood, and 1,500–3,500 kg of melons. Market products came partly from the peasants' own plots and the fields newly opened by city units, but the city still had to continue the fight against trade in rationing tickets.[433]

In May, the City Government decided again on material rewards for 1962 for selling grain and vegetables voluntarily to the state. Barter trade was expanded through handing out purchasing certificates for daily necessities such as aluminium jars, thermos flasks, radios, toothpaste, electric torches, locks, cloves, food tables, etc. To encourage opening new fields, every new *mu* was given 20 *jin* in grain reward. For earthwork projects, every square metre finished was rewarded with 1 *liang* of grain. It was emphasized that no rewards were to be given to those who were not energetically opening up new fields, were lax on management, or were engaged in the private handling of commercial vegetables.[434] Also, cotton, edible oils, hemp, and animals and poultry reared by individuals were to be given material awards.[435]

As a result of these policies, vegetable growing continued expanding notably. However, much of the newly reclaimed land was too saline and did not produce well at all. The policy also brought with it the problem of losing dry fields from grain to growing vegetables.[436] When supplies got better in 1963, the most unsuitable vegetable fields were decommissioned and 5,000 hectares returned to other uses.[437] The drive to reduce the urban population continued throughout 1962. At the end of August, the City Government issued instructions on yet another campaign to check and verify the urban population.[438]

In September 1962, the supply situation was getting visibly better, which could be seen in the gradual relaxation of rationing. When the autumn harvest began in late September, the voucher system on supplementary foodstuff was ended. But the City Bulletin still reminded people that the production of many important products, such as grain, edible oils, meat, vegetables and cloth was still insufficient. Therefore, the campaign to economize and promote production had to continue, and people were not to grow lax in this respect.[439]

In August, the Party Central Committee held its normal summer meeting in Beidaihe. At this meeting, Chairman Mao attacked the trend of what he saw as *de facto* return to private agriculture and reasserted the role of collectivism and people's communes as the backbone of socialist agriculture. He also warned the gathered communist leaders of the possibility of revisionism in the Party and reminded them of continuous class struggle. After the meeting, the Party Centre started the Socialist Education Campaign in the countryside aimed at teaching peasants a proper socialist outlook. This campaign continued until the Cultural Revolution and was naturally also undertaken in Tianjin.

The most visible sign of the tightening political atmosphere was the reversal of the city policy on peddler markets in the latter half of 1962. At the end of August, the City Government issued an order on strengthening vegetable sales controls and prohibiting collectives from selling their produce directly to *danweis*. The order complained about production brigades and individual peddlers' activities of bypassing state procurement, which had to cease. Production brigades were threatened with deductions to their grain rations if they did not stop their activities and peddlers were once again banned from going to suburbs to buy vegetables on their own. Instead, they had to get their vegetables from state wholesalers, if they had any goods to sell to them.[440]

In November, further restrictions were decided on peddlers and their markets.[441] The decision was grounded on the fears that the earlier policy was as too capitalist. As a consequence, free markets were closed down gradually in the city. A case where a young girl exposed her mother for peddling as 'removing foundations from the wall of socialism' was propagated throughout the city.[442] As a part of the tightening policy atmosphere, new Supplementary Food Production Management Committees were established in *xitongs* and companies to exercise control over their supplementary food production and to curb their engagement in black markets for vegetables.[443] This was justified by Mao's Beidaihe remarks on taking 'class struggle as the key'.[444]

Population reduction policies also continued in 1962. As late as May 1962, the reduction target for the population was expanded from 240,000 to 400,000.[445] The policy was not so popular in many *danweis*, which needed manpower, and, needless to say, many of their temporary peasant workers did not want return to their villages. However, in November, the City Government reaffirmed the policy and noted that all decisions on hiring people had to be ratified by the City Planning Commission and Labour Bureau following central, provincial and city guidelines. Labour Bureaus and Management Bureaus had to investigate the cases of *danweis* hiring people without approval and deal with such cases strictly.[446] As Brown has shown, in the end a total of 222,064 people were sent home from Tianjin.[447]

Assessment of Tianjin Famine Managment

A body count is an inevitable part of a thesis on a famine. The author could not unearth any internal report on the total number of 'unnatural' deaths in Tianjin 1959–61, which does not mean that one does not exist somewhere. However, literature offers many estimates of the numbers of starved and killed people in China and its provinces during the GLF, including Tianjin. Based on official population statistics, Yang Jisheng has recently come up with 32,600 unnatural deaths and an 87,600 shortfall in births in Tianjin during this period,[448] but he does not provide a breakdown of figures for urban and rural residents, or how the constant changes in administrative borders affected the outcome. It may be necessary also to include some deaths in Hebei in the figure as well.

However, using Tianjin population statistics we can compute an estimate of urban casualties. Although the statistics naturally do not give the exact causes of deaths per year, they do allow us to compare death rates in the city before and during the GLF. As for urban *hukou*-holders, the average death rate of the previous seven years (1952–8) was 7.74 per cent, and for the previous three years (1956–8) 6.73 per cent. The average death rate during the three famine years in the city was 7.94 per cent. The differences between the average death rates were therefore 0.2 per cent (the seven years average minus the GLF average) or 1.21 per cent (the three years average minus the GLF average), depending on which figure you use for comparison. If we assume that the difference between the earlier rates and the GLF rate was all caused by the famine, then depending on the way we compute it, the famine killed anything between 619 and 3,744 city *hukou*-holders.[449] Compared to Yang Jisheng's figure above, these figures would indicate that rural areas suffered about 28,000–32,000 deaths at the same time.

It can be noted that before the GLF, and then again after 1961, death rates were dropping constantly in the city, and the GLF made a clear exception to this general trend, therefore at least part of the 'excess' deaths in urban areas were very likely to be caused by the famine. Other causes could have included, for example, industrial accidents which rose during the construction push. At the same time, we cannot trust the statistics, since sincere reporting of fatalities was not on the party-state agenda. Officials did not keep exact accounts of the people who starved during the famine and their causes of death – why would they have recorded their own crimes in the first place? Moreover, residents themselves could for example hide their dead relatives in order to get their rations. The figure is, as all figures concerning the GLG famine death rates, therefore only a rough estimate.

The legacy of the famine was complex in Tianjin: the urban areas averted a full-blown disaster, but suffered three years of food crisis and dearth. Rural areas became disaster zones due to central and city policies. It is clear that the city could have done more for its rural areas from the beginning. It could have been more lax on its policies, it could have concentrated on food production more, and it could have cut urban rations in favour of peasants and allowed peasants to rest more, like it did with urbanites. However, when compared to many other areas in China, the Tianjin countryside was relatively lightly affected. This may have been a result of Tianjin being a grain production deficit area to begin with and thus being subject to fewer demands for grain. Moreover, the city was a place of refuge for many who managed to get in and a majority of these people came from Tianjin's own countryside. How much the interventionist disaster management style of the city helped in the outcome is difficult to say, but it is evident that Maoist disaster management practices as they were practiced in Tianjin did increase peasant vulnerability to famine in general.

Also the tension between self-help and social-help was clear in Maoist disaster management style during the famine. The leaking food supply system was basically a sign that everybody, both individuals and work units, put their needs above the collective needs and did not trust that the city could provide them with their entitlements to food. Subverting the monopoly food procurement and distribution system was everyone's game. It started in the villages, where grain was concealed and vegetables and other foodstuff sold to private peddlers or work units for higher profits and commodities. It continued in the supply chain, where everybody skimmed the produce, especially the cadres in charge of allocation work, and continued to *danweis*, which constantly sought to 'cooperate' with suburban food producers to guarantee their own food supplies.

In a sense, the famine exposed the state monopoly system to a prisoner's dilemma problem it could not solve. People and units could not just sit and wait to get their share promised to them while others took things into their own hands and helped themselves. Disciplinary measures and moral persuasion, organizational rearrangements, etc. did not work, which is attested to by the fact that the city had to order the same things over and again and be under a constant 'checking and verification' drive from early 1959 to 1962.

The Tianjin experience during the Great Leap Forward famine also shows a number of the limits and strengths of Maoist campaign-based disaster management practices. First, it shows how the central government's role was crucial in local disaster management, despite the fact that relying on local initiative and autarky was emphasized in the official rhetoric. First of all, the city needed central approval to fully use its own means and resources for disaster management which was always a distraction from the city's regular activities, especially its production duties. This led to the question of what campaigns to prioritize and, in the end, such decisions could only be made by the central government. It took until August 1960 for the Party Centre to allow the reallocation of substantial resources away from economic construction to mitigate the crisis. Before this the city could only engage in disaster management activities that did not interfere with production. Outright from directly disobeying central authorities, the city's hands were tied. This way the GLF famine exposed the weaknesses of the Maoist disaster management style. Politics were in command also in Maoist disaster management.

Notes

1 See calculations of mortality rates in different localities in Peng (1987).
2 TJTongji 1999, 31.
3 GCDTJ 1991, 130; TJTZHZHQ 1996, 5; Lieberthal 1980 on the establishment of Communist rule in Tianjin.
4 LSJ / GB 1994, 4.
5 Ibid.; TJPrice 1997, 66.
6 Mao Zedong (Tse-tung) (1955a).
7 Mao Zedong (Tse-tung) (1955b).
8 The First and Second Commercial Bureaus were in charge of managing retail of other non-edible commodities circulating in the city economy. In September 1959 the Third Commercial Bureau was reorganized and renamed the City Supplementary Food Office, but its jurisdiction remained the same.
9 The classification system for grains also included some other foodstuff that strictly speaking does not correspond to the English translation of grain (*liang*), as it included beets and roots such as potatoes, sweet potatoes and others.
10 LSJ / GB 1994, 172 and 174–5.
11 Xie 2011.
12 LSJ / GB 1994, 146–7 and 155.
13 Cheng and Selden 1999.
14 LSJ / GB 1994, 111.
15 Ibid., 112–114.
16 Ibid., 112 and 132–3.
17 Ibid., 5.
18 Ibid., 132–3.
19 Ibid., 118–19.
20 Ibid., 114.
21 ES 2005, 470–81.
22 Since 1959 1 *liang* has been equivalent to 50 g, before this the old *liang* equalled 31.5 g.
23 GS 2001, 123–4.
24 Ibid., 120–1.
25 LSJ / GB 1994, 119–20.
26 GS 2001, 121; TJPrice 1997, 26.
27 GS 2001, 21.
28 Ibid., 122–3.
29 Brown 2012, 61.
30 TJZL 14 1982, 30; Guo 1995, 149.
31 LSJ / GB 1994, 5–6; TJJZ 1991, 5–6.
32 GCDTJ 1991, 147; LSJ / GB 1994, 23 and 91–2.
33 GCDTJ 1991, 146–7.
34 TJZL 14 1982, 31.
35 Guo 1995, 160.
36 The counties were: Yanshan, Huanghua, Jinhai, Renqiu, Wuqing, Ba, Hejian, Jiaohe, Xian, Wuqiao and Cang.
37 TJTongji 1999, 3–4.
38 Brown 2012, 234.
39 Zhou 2012, 11.
40 TJTongji 1999, 3–4.
41 GCDTJ 1991, 13; DDZHGTJ Vol. 1 1989, 120.
42 LSJ / GB 1994, 81.
43 The beginning of the GLF is described in MacFarquhar (1983) and Teiwes and Sun 1999.
44 TR 18.2.1958, 1.
45 Guo 1995, 148–50; DDZHGTJ Vol. 1 1989, 108–9; TJZL 14 1982, 33; CJConst 1987, 249–50.

46 CJConst 1987, 249–50.
47 TJZL 14 1982, 29.
48 Guo 1995, 148 and 152; TJZL 14 1982, 34–5.
49 TJZHB 21 (26.5.1958), 6–7.
50 Guo 1995, 148.
51 Becker (1996) describes the new 'scientific' agricultural methods used in the experimental fields and their failure as well.
52 TJZL 14 1982, 14.
53 TJZHB 39 (28.9.1959), 4.
54 TJZL 14 1982, 42; TJTZHZHQ 1996, 243.
55 DDZHGTJ Vol. 1 1989, 120.
56 CJConst 1987, 251.
57 Guo 1995, 155.
58 Ibid., 157.
59 DDZHGTJ Vol. 1 1989, 120.
60 TJZL 14 1982, 45–6; DDZHGTJ Vol. 1 1989, 119.
61 Guo 1995, 161.
62 Ibid., 159; TJTZHZHQ 1996, 57.
63 DDZHGTJ Vol. 1 1989, 120.
64 Ibid., 108–9; TJDS 1995, 362.
65 TR 1.10.1958, 6.
66 LSJ / GB 1994, 146–7.
67 TJZL 14 1982, 42.
68 Guo 1995, 159.
69 TR 11.9.1958, 1–2
70 TJDS 1995, 36.
71 CJConst 1987, 251; however, including the countryside this figure rose to 24,845 small furnaces (Guo 1995, 161).
72 DDZHGTJ Vol. 1 1989, 112.
73 Ibid., 111–12.
74 CJConst 1987, 251.
75 LSJ / GB 1994, 114 and 75.
76 TCA X0096-C-000026-012 (1.7.1959).
77 Guo 1995, 164–5.
78 Brown 2012, 58.
79 TJZL 14 1982, 53.
80 Ibid., 50.
81 Guo 1995, 163.
82 TJZHB 1 (5.1.1959), 10.
83 GCDTJ 1991, 148.
84 TJZL 14 1982, 53; CJConst 1987, 317.
85 CJConst 1987, 251–2.
86 TJZL 14 1982, 73.
87 Ibid., 57; Guo 1995,167.
88 TJZL 14 1982, 14, 57–8; Guo 1995, 167.
89 LSJ / GB 1994, 4 and 175.
90 TJZHB 7 (16.2.1959), 9.
91 LSJ / GB 1994, 155.
92 Becker 1996, 86–7; Bernstein 1984, 361–2; Dikötter 2010, 80–2 and 88 on the spring of 1959.
93 TJTZHZHQ 1996, 59.
94 TCA X0283-c-000868-069 (1.6.1959), 1–2.
95 Yang 2009, 278.
96 TCA X0283-c-000868-069 (1.6.1959), 2–3.

97 Ibid., 3–4.
98 Ibid.
99 Yang 2009, 278.
100 Yang 2012, 216–19.
101 Xie 2011; according to Yang (2009, 278) over 80 per cent of the affected counties' people were suffering from grain shortages.
102 Brown 2012, 62–4.
103 TJZL 14 1982, 68.
104 TCA X0095-C-000754-048 (1.9.1959).
105 TCA X0095-C-000754-048 (May 1959).
106 TCA X0095-C-000754-048, (22.5.1959), 1–2.
107 Ibid.
108 TCA X0095-C-000754-048 (May 1959).
109 TJZHB 15 (11.4.1960c), 12–13.
110 Ibid.
111 TJZHB 15 (11.4.1960f), 14.
112 Ibid.
113 TJZHB 15 (11.4.1960d), 13–14.
114 Such regulations had existed since 1954 and the 1959 campaign was technically about reinstating these regulations (GS 2001, 113–114).
115 Brown 2012, 41
116 GS 2001, 113 and 129–30; TJZHB 24 (16.6.1958), 13–14; TCA X0095-C-000754-048 (May 1959).
117 TJZHB 45 (10.11.1958), 9; TJZHB 7 (16.2.1959), 9.
118 TJZHB 45 (10.11.1958), 9.
119 TJZHB 7 (16.2.1959), 9.
120 TJZHB 18 (4.5.1959), 2–3.
121 TCA X0096-C-000026-012 (15.6.1959), 13.
122 TJZHB 26 (29.6.1959), 3.
123 TJZHB 23 (8.6.1959), 10.
124 TJDS 1995, 363–4.
125 TJZHB 14 (6.4.1959), 9–10.
126 TCA X0095-C-000754-04 (22.5.1959).
127 Guo 1995, 167.
128 TJZHB 11 (16.3.1959), 8–9.
129 LSJ / GB 1994, 234.
130 TJZHB 28 (13.7.1959), 3.
131 TJZHB 22 (1.6.1959), 2–4.
132 TJZHB 24 (15.6.1959), 2–3.
133 Teiwes and Sun 1999, 162–4; Dikötter 2010, Chapter 11.
134 Guo 1995, 169.
135 TJZL 14 1982, 60.
136 TJZHB 24 (15.6.1959), 2–3.
137 TCA X0096-C-000026-012 (15.6.1959).
138 Guo 1995, 173.
139 TCA X0096-C-000026-012 (15.6.1959).
140 Guo 1995, 173.
141 *Zhong-Gong zhongyang* (4.7.1959).
142 TJZL 14 1982, 73.
143 TJDS 1995, 369.
144 Guo 1995, 173 and 180.
145 TR 20.9.1959a, 1.
146 TR 20.9.1959b, 1.
147 TJZHB 30 (27.7.1959), 6–12.

148 TR 27.8.1959, 2
149 TCA X0095–C-000754-048 (1.9.1959).
150 TCA X0196-C-001086-037 (10.10.1959).
151 TJZHB 15 (11.4.1960f), 14.
152 TJZHB 15 (11.4.1960a), 8–9.
153 TJZHB 36 (7.9.1959), 7.
154 GS 2001, 127–8 and 131–3.
155 MZHJZH 1995, 30.
156 Brown 2012, 44.
157 MZHJZH 1995, 291–5.
158 Ibid.
159 TJZHB 33 (17.8.1959), 5–7.
160 TCA X0196-C-001086-037 (10.10.1959).
161 TJZL 14 1982, 75 and 79.
162 TCA X0196-C-001086-037 (10.10.1959); Guo 1995, 178.
163 LSJ / GB 1994, 147.
164 On this, see Thaxton 2008, 244.
165 TJZL 14 1982, 83.
166 Ibid.
167 Yang 2009, 282.
168 TR 14.10.1959, 1; TR 22.11.1959, 3.
169 TJZHB 52 (28.12.1959), 5–7.
170 TJDS 1995, 363–4; TJTZHZHQ 1996, 59.
171 TJZHB 45, (9.11.1959a), 8–9.
172 TR 19.11.1959, 1
173 TJZHB 27 (4.7.1960a), 2–5.
174 TJZHB 45 (9.11.1959b), 14.
175 DDZHGTJ Vol. 1 1989, 122–3.
176 TCA X0095-C-000754-048 (25.12.1959); TCA X0096-C-000026-012 (15.11.1959).
177 TJZHB 2 (11.1.1960), 4.
178 TR 31.12.1959, 1.
179 TJZL 14 1982, 86; Guo 1995, 183.
180 Bernstein 2006.
181 Guo 1995, 180.
182 TJZHB 4 (25.1.1960), 3.
183 Guo 1995, 184.
184 TR 17.1.1960, 1.
185 DDZHGTJ Vol. 1 1989, 123.
186 TJZHB 17 (25.4.1960), 6–7; TR 17.1.1960, 1.
187 TR 26.7.1960, 2.
188 TJZL 14 1982, 108.
189 TJZHB 15 (11.4.1960b), 11–12.
190 TJHealth 1992, 59.
191 TCA X0053-C-001746-003 (7.3.1960).
192 Yang 2009, 279.
193 TCA X0053-C-001746-003 (7.3.1960).
194 Ibid.
195 Wang 2001, 124–5.
196 TCA X0003-Y-009036 (5.5.1960).
197 Guo 1995, 184.
198 TCA X0003-Y-009036 (22.4.1960); TCA X0003-Y-009036 (27.4.1960).
199 Yang 2009, 281.
200 Wuqing xianzhi 1991, 44–5.
201 TJPrice 1997, 27.

202 TJZL 14 1982, 89.
203 LSJ / GB 1994, 156.
204 The length of such meetings were often prolonged to give cadres a chance to improve their personal nutrition situation through being able to enjoy better rations for the duration of meetings. Guo 1995, 182; TJZL 14 1982, 90.
205 TJZL 14 1982, 89.
206 LSJ / GB 1994, 229.
207 TJZHB 19 (9.5.1960), 6–7.
208 TJZHB 27 (4.7.1960b), 9–10; reports on the campaign: TJZHB 18 (2.5.1960), 11–12.
209 TJZHB 15 (11.4.1960b), 11–12.
210 TJZHB 15 (11.4.1960a), 8–9.
211 LSJ / GB 1994, 164.
212 Yang 2009, 279.
213 TCA X0003-Y-008983 (16.1.1960), 1–2.
214 TJZHB 15 (11.4.1960a), 9.
215 TJZHB 32 (8.8.1960a), 5.
216 MZHJZH 1995, 293.
217 TJZHB 32 (8.8.1960a), 4–5.
218 TJZHB 15 (11.4.1960f), 14; TJZHB 15 (11.4.1960c), 12–13.
219 TCA 401206800-X0085-Y-000055-016 (30.5.1960).
220 Gao 2011, 173.
221 TJZHB 32 (8.8.1960a), 4–5.
222 TCA X0096-C-000026-012 (15.6.1959), 1–2.
223 LSJ / GB 1994, 175
224 Brown 2011, 236.
225 Xie 2011, 6–7.
226 TCA X0095-Y-000753-038 (28.6.1960).
227 TJZHB 15 (11.4.1960b), 11–12.
228 TJDS 1995, 374.
229 Ash 2006, 986.
230 Guo 1995, 187; TJZL 14 1982, 98.
231 TJZHB 32 (8.8.1960a), 4–5.
232 Ibid.
233 TJZHB 35 (9.29.1960), 10.
234 TR 19.7.1960, 1
235 Guo 1995, 188.
236 TJZHB 30 (25.7.1960), 8–9.
237 TJZHB 32 (8.8.1960b), 6–8.
238 Wan Xiaotang 9.11.1960/2001, 274–6.
239 Guo 1995, 188; TJZHB 32 (8.8.1960b), 6–8.
240 LSJ / GB 1994, 156 and 164.
241 ES 2005, 20.
242 TJZHB 43 (7.11.1960), 12.
243 TJZHB 32 (8.8.1960b), 7.
244 Ibid.
245 These were criticized by Vice-Mayor Song Jingyi in a cadre vegetable-growing meeting in March 1961: TJZHB 6 (31.3.1961), 2–5.
246 TJZHB 26 (27.6.1960), 8–9.
247 TJZHB 27 (4.7.1960), 10.
248 TJZHB 27 (4.7.1960c), 8–10.
249 TJZHB 32 (8.8.1960c), 6.
250 DDZHGTJ Vol. 1 1989, 123–4.
251 TJZHB 45 (5.12.1960c), 6–7.
252 Gao 2011.

253 Guo 1995, 187 and 189.
254 TJZL 14 1982, 14, 102; LSJ / GB 1994, 234.
255 TJDS 1995, 374; TJZL 14 1982, 102; LSJ / GB 1994, 234.
256 LSJ / GB, 1994, 119; Guo 1995, 186.
257 Tao 2001, 170–1.
258 TJZHB 32 (8.8.1960a), 4.
259 Yang 2009, 285.
260 Li Ding, 2001, 47–8.
261 Dakang 1994, 14; TR 10.9.1960, 4.
262 TJZL 14 1982, 100.
263 TR 20.8.1960, 2.
264 TR 27.8.1959, 2.
265 TJZHB 30 (25.7.1960), 8–9.
266 TJZHB 32 (8.8.1960a), 4–5.
267 TJZHB 37 (12.9.1960), 2–5
268 Guo Zhiying 1995, 190; TJZL 14 1982, 102.
269 TJZHB 40 (3.10.1960a), 2–3
270 TJZHB 37 (12.9.1960), 2–5.
271 TJZHB 37 (12.9.1960), 3–4.
272 TCA 401206800-X0085-Y-000055-016 (September 1960).
273 TCA X0003-Y-008983 (16.10.1960).
274 TCA X0003-Y-008983/28 (5.10.1960).
275 MZHJZH 1995, 291–5.
276 TCA 401206800-X0085-Y-000055-016 (September 1960); TJZL 14 1982, 14, 102; LSJ / GB 1994, 234.
277 TJZHB 40 (3.10.1960b), 2.
278 TJZL 14 1982 14, 103; inflows of temporary peasant workers into Tianjin have been documented in Brown 2012, 55–6.
279 TJZL 14 1982, 103–4.
280 TJZHB 45 (5.12.1960b), 10.
281 Guo 1995, 191; TJZL 14 1982, 104; TJDS 1995, 374–5.
282 Tao 2001, 158–84, 170.
283 TJZHB 44 (21.11.1960b), 3–4.
284 TJZHB 44 (21.11. 1960a), 2.
285 Li Ding 2001, 47–8.
286 Wan 1960/2001, 274–6.
287 Tao 2001, 170–1.
288 Wan 1960/2001, 274.
289 Ibid.; TJZL 14 1982, 107; TJDS 1995, 375.
290 Tao 2001, 170.
291 *Jinghai Xianzhi* 1995, 21.
292 *Hangu Quzhi* 1995, 28.
293 TJZL 14 1982, 108–9.
294 TJZL 14 1982, 106–7; DDZHGTJ Vol. 1 1989, 123.
295 *Tianjin weisheng daji* 1992, 62 and 68.
296 TJZHB 45 (5.12.1960a), 6–7.
297 TJZHB 45 (5.12.1960a), 7; TR 30.11.1960, 1.
298 TR 30.11.1960, 1
299 TCA X0095-C-000793-016 (10.12.1960); TCA X0095-C-000793-016 (24.12.1960); Guo Zhiying 1995, 195.
300 Xie 2011.
301 TCA X0095-C-000793-016 (24.12.1960).
302 TCA X0196-C-001086-037 (22.12.1961).
303 Xie (2011) gives a figure of 900,000 oedema patients in the city.

304 TCA X0095-C-000793-016 (24.12.1960).
305 TCA X0095-Y-000753-038 (20.12.1960).
306 TCA X0095-C-000793-016 (28.11.1960).
307 TCA X0196-C-001086-037 (18.1.1961).
308 TCA X0196-C-001086-004/8-15; TCA X0196-C-001086-004/18 (10.1.1961)
309 TCA X0196-C-001086-004/21 (14.4.1961), 6.
310 TCA X0095-C-000793-016 (24.12.1960); Xie 2011.
311 TCA X0196-C-001086-004/38 (22.1.1961), 3.
312 TCA X0196-C-001086-037 (18.1.1961).
313 Yang 2012, 216–19.
314 TCA X0198-Y-000984-002 (28.2.1961).
315 TCA X0003-C-010770/29 (4.2.1961).
316 TCA X0196-C-001086-004/38 (22.1.1961), 3–4.
317 TCA X0198-Y-000984-002 (28.2.1961).
318 Ibid.
319 TCA X0196-C-001086-004/21 (14.4.1961).
320 Guo 1995, 192; TJZL 14 1982, 106.
321 DDZHGTJ Vol. 1 1989, 123.
322 TCA X0196-C-001086-004/21 (14.4.1961).
323 Ibid.
324 TJZL 14 1982, 108.
325 Shenpanzhi 1999, 211
326 Xie 2011.
327 TCA X0196-C-001086-037 (22.12.1960); Cooper and Liu 2002, 283–5, 293.
328 TCA X0196-C-001086-004/21 (14.4.1961).
329 Li Zhongyuan 2001, 40.
330 TCA X0196-C-001086-004/35 (29.11.1961).
331 TJZHB 45 (5.12.1960d), 9–10.
332 TJZHB 1 (17.1.1961a), 2–5.
333 Brown 2012, 59.
334 TCA X0003-Y-008977 (19.12.1960).
335 Ibid.
336 Wang 2009.
337 Tao 2001, 172–3.
338 Brown 2012, 64.
339 Shenpanzhi 1999; Brown 2012, 64.
340 These were: 'readjustment, consolidation, filling out and raising standards' of the economy.
341 TJTZHZHQ 1996, 244.
342 Xie 2011.
343 Tao 2001, 170–1.
344 LSJ / GB 1994, 114,119 and 189; TJZL 14 1982, 111.
345 Guo 1995, 199.
346 TCA X0095-C-000793-016 (28.4.1961).
347 Li Ding 2001, 49.
348 TJPrice 1997, 28.
349 TJZL 14 1982, 112–13.
350 TJDS 1995, 376.
351 TJZHB 5 (22.3.1961), 5–7.
352 TR 30.3.1961, 2.
353 LSJ / GB 1994, 121.
354 Wang 2009.
355 Tao 2001, 172–3 .
356 TCA X003-C-010877/28 (4.2.1961).

357 TCA X003-C-010877 (13.2.1961).
358 TCA X028-C-001101-023/94 (16.5.1961).
359 Yang 2009, 281–2.
360 TJZHB 1 (17.1.1961b), 6.
361 ES 2005, 5.
362 TCA X0196-C-001086-037 (27.3.1961).
363 TJlaogai 1999, 25.
364 TR 12.2.1961, 1.
365 TR 22.2.1961, 1.
366 Ibid.
367 TJZL 14 1982, 115; Guo 1995, 198; TJDS 1995, 377.
368 TJZHB 6 (31.3.1961), 2–5.
369 TR 10.3.1961, 1.
370 TR 14.3.1961, 2; TR 21.3.1961, 2.
371 TJZL 14 1982, 116.
372 Friedman *et al.* 1991, 259–60.
373 TJZHB 20 (31.10.1961b), 6.
374 Guo 1995, 206–7; TJZL 14 1982, 131.
375 Some material rewards for production in reclaimed land were introduced already in late 1960, but this became systematic in 1961. Already at the end of 1960 it had been decided that for each newly opened *mu* of field, 20 *jin* of 'reward grain' would be given (LSJ / GB 1994, 164).
376 TJZL 14 1982, 112 and 121.
377 DDZHGTJ Vol. 1 1989, 123.
378 TJZL 14 1982, 124.
379 TR 17.6.1961, 1.
380 TJZL 14 1982, 125.
381 Ibid., 121–3.
382 Ibid., 121–123.
383 Guo 1995, 204.
384 LSJ / GB 1994, 114.
385 TJZHB 11–12 (30.6.1961a), 2–3.
386 TJZL 14 1982, 123.
387 Li Zhongyuan 2001, 39–40.
388 TJZHB 24 (25.12.1961), 3.
389 TJZHB 19 (20.10.1961), 2–4.
390 TJZHB 24 (25.12.1961), 3.
391 TJZHB 19 (20.10.1961), 2–4.
392 TJZHB 19 (20.10.1961), 4–5.
393 LSJ / GB 1994, 120; TJZHB 19 (20.10.1961), 4–5.
394 TR 12.8.1961, 2.
395 TJZHB 20 (31.10.1961a), 3.
396 TJZHB 24 (25.12.1961), 2–4.
397 LSJ / GB 1994, 156 and 164; ES 2005, 477.
398 TJZL 14 1982, 131.
399 TJZHB 21–22 (30.11.1961), 2.
400 Ibid., 2–3.
401 LSJ / GB 1994, 121 and 167–8.
402 ES 2005, 477.
403 LSJ / GB 1994, 164.
404 *Jinnan quzhi* 1999, 33.
405 TJDS 1995, 381.
406 *Jinnan quzhi* 1999, 33.
407 GS 2001, 133.

408 Ibid.
409 TJPrice 1997, 6.
410 TJZHB 11–12 (30.6.1961b), 8–9; and GS 2001, 133–6.
411 GS 2001, 133–6, GS 2001, 135.
412 TJZHB 1 (12.1.1962), 7.
413 Ibid.
414 TJDS 1995, 379.
415 Guo 1995, 208–9; TJZL 14 1982, 136.
416 Shenpanzhi 1999, 29.
417 Guo 1995, 207.
418 TJZL 14 1982, 134.
419 LSJ / GB 1994, 175.
420 TJZHB 3 (10.2.1962), 8.
421 TCA X0196-C-001086-004/35 (2.12.1961).
422 Guo 1995, 209.
423 TJTZHZHQ 1996, 244–5.
424 See MacFarquhar (1997, 142–79) on the meeting.
425 GCDTJ 1991, 131.
426 Guo 1995, 211–12.
427 Here figures in different sources differ probably due to area mergers in Tianjin.
428 DDZHGTJ Vol. 1 1989, 125–6.
429 LSJ / GB 1994, 176.
430 TJHealth 1992, 8.
431 Ibid., 8 and 68.
432 TR 25.1.1962, 1.
433 GS 2005, 137–8.
434 TJZHB 8–9 (12.5.1962), 6–8
435 TJZHB 16 (31.8.1962), 2–3
436 LSJ / GB 1994, 157–9 and 164–5.
437 ES 2005, 21.
438 TJZHB 16 (31.8.1962b), 4.
439 TJZHB 18 (15.10.1962), 4–5.
440 TJZHB 16 (31.8.1962c), 9.
441 Guo 1995, 222–3.
442 DDZHGTJ Vol. 1 1989, 135–137
443 Guo Zhiying 1995, 223
444 DDZHGTJ Vol. 1 1989, 135–6.
445 TJDS 1995, 382.
446 TJZHB 22 (26.11.1962), 4.
447 Brown 2011, 246.
448 Yang 2009, 414.
449 The figures are from *Tianjin Renkou* (1986, 4–10); the estimates are computed as follows: the actual average number of dead 1959–61 - city average population 1959–61 × the difference between the average death rates before the GLF and during the GLF (computed based on both the 3 and 7-year averages) = number of 'extra dead' in 1959–61. In the source, the figures for counties are given only for counties under Tianjin jurisdiction in 1986, not in 1959–61, so they cannot be used for calculations for rural death rates. Also the borders of urban districts may have changed after the GLF, but their key areas have remained the same.

2 The Great Flood of 1963

By 1963 the great famine was over. But, as nature would have it, Tianjin was to face another crisis that threatened to turn into a large-scale disaster. This time it was about unusually strong monsoon rains that triggered devastating floods in Western and Southern parts of Hebei Province. Tianjin was in the path of the main discharge waters of the flood, and therefore the city now faced, right after starvation, the danger of being inundated. The management of the flood was in many ways different to the just-finished famine. It consisted of pre-planned measures, long and short-term disaster control preparations and the quick mobilization of nearly 1 million city residents to save the city. There was also a systematic relief aid effort directed at flood refugees and reconstruction of flood-hit areas after the waters finally receded. Tianjin city rallied in a way it had not done during the GLF famine, and the great flood of 1963 can be said to have been one of the major examples of successful disaster management during the Maoist era. Paradoxically, it was the same Maoist socio-political structures that had made famine management so difficult that helped to save the city in 1963.

Disaster Preparedness

Preparing for future disasters is an important part of disaster management. It can determine the outcome between success and failure. Some types of natural disasters are more predictable than others. Floods, earthquakes and volcanic eruptions always tend to happen at least roughly speaking at the same locations. However, in between the hazards these places can be attractive locations for establishing human settlements, even large cities, such as Tianjin. Trusting their luck, gods or technology, and often having too few alternatives, people build their cities in such natural disaster-prone areas everywhere in the world. China is no exception to this. As discussed in the Introduction, 'taming the waters' around the frequently flooding Yellow River was mentioned as one of the founding acts by sage Yu in the creation of the first unified Chinese state. As much as it is a legend, it is also about the reality of floods as a constant feature in the political and social ecology of China, a fact of nature to be reckoned with, and one that also the communist regime encountered many times.

Protecting human settlements from floods has an old tradition in China and has been regarded as a key duty of the government as discussed above. In many places in China, rivers are trained by directing their flows into channels that can even run higher than the surrounding farmlands. Elaborate systems of discharge channels, sluices and reservoirs are used to guide waters safely past human settlements and to irrigate fields, and, in more modern times, for power generation. This also has been the general situation in the Hebei plains where many rivers have been directed to run above the level of the surrounding farmland. They also carry a lot of silt from the Taihang uplands, which means they require constant maintenance and dredging.[1] Tellingly, during the early Qing period the Northern, and the biggest, tributary river to the Hai He, the Yongding River was called the Wuding He or the 'River with No Fixed Course', because of the constant floods and changes in its channel. After flood control measures were taken at that time on the river, its name was changed into Yongding He or the 'River with a Permanently Fixed Course'.[2]

Not always have flood protection measures worked in Hebei. In 1939, Tianjin was submerged for a month. In 1954, 1956, 1961 and 1962 there were heavy rains causing local damage to production and property in the city. However, after the establishment of the PRC, the communist regime took a number of measures to protect the city better. In 1953, a large discharge channel called Duliujian River was constructed in the countryside southwest of the city. It had almost the same discharge capacity as the Hai He.[3] Protecting Duliujian embankments became one of the key sites of anti-flood struggle in 1963.

The construction of irrigation and flood protection waterworks was also a major part of the Great Leap Forward. Indeed, the name of the campaign originally referred to the large waterworks taken in the winter of 1957–8 all over China. A promise of better irrigation systems and protection from natural disasters was one of the justifications for the establishment of the people's communes. On paper at least, the construction projects in Hebei Province during the GLF were impressive. From 1958, the province constructed 17 large, 19 middle-sized, and nearly 1,000 small reservoirs.[4] In 1963, according to the Provincial Flood Protection Work Plan for 1963, 80 per cent of the Haihe drainage area was officially under various 'controls', the river system's discharge capacity had been increased and management improved. However, the Plan also noted that large and middle-sized reservoirs had not experienced a great flood, and in some of them construction criteria had been low, many discharge channels were small in their capacity, many embankments had gone uninspected, and especially in tributary rivers many embankments were unfinished. Manpower shortages slowed down fixing these problems.[5]

In 1963, long-term weather forecasts indicated that the year would be abnormally rainy. This was emphasized both by the Central Flood Protection Headquarters and the Hebei Provincial Flood Protection Headquarters (HBFPHQ) in their instructions to localities issued in late May and June respectively.[6] They stressed the need to inspect, clean and repair reservoir walls and river embankments and establish organizational readiness for flood protection work. Furthermore, evacuation personnel and work in water diversion areas had to be organized in good time. To organize this work, the HBFPHQ held an all-province flood protection meeting in late May – early June with Tianjin representatives participating. The meeting noted that, since May, rainfall had been abnormally high in the province and people should prepare for a 1956-like flooding, which had been the severest flood since the Liberation. The meeting called for carrying out propaganda work, and improving material preparations and organizational work with the general aim that reservoirs would not leak and embankments would not collapse. Another aim was to protect industrial and agricultural production, which was summarized in a slogan 'big flood will cut production only a little, small flood only increases crops'.[7]

The meeting further suggested that there should be a province-wide rush to repair infrastructure and finish the existing construction projects by the end of June. Leading cadres responsible for flood protection work were to inspect infrastructures and report the results to the province by the end of June. All reservoirs and sections of embankments were to have responsible cadres for inspections and repairs. The masses were to be organized into stand-by troops, rush repair troops and reserve troops and assigned their duties beforehand on different levels.[8]

The Provincial Flood Protection Work Plan for 1963 further defined the flood protection tasks for cities noting that, in the case of a big flooding, cities that could be kept [i.e. not allowed to be inundated] had to be kept at all cost. Cities should mobilize masses to support disaster areas, and if hit themselves, their people had to be promptly 'consoled', after which production had to be resumed as soon as possible. Tianjin was ordered to immediately mobilize and take 'all kinds of measures' [sic] to protect the city. Cadres and party members there, as everywhere in the province, had to go to the frontline to take the lead in the struggle, swallowing the bitter pill with the masses.[9]

The Tianjin City Government followed the instructions and established Tianjin City Flood Prevention Headquarters (*Tianjinshi fangxun zhihuibu*, hereinafter TJFPHQ) on 15 June 1963,[10] with Vice-Mayor Wang Peiren as the chair.[11] As one of its first acts, the headquarters drafted the Tianjin City Flood Protection Work Plan. It showed how bureaucratic routine work laid the basis for emergency work in the city. Under the TJFPHQ, the City Construction and the Hydro Power Bureaus were instructed to set up two subdivisions for city and suburban districts alike to administer flood prevention work in these areas. In the suburbs, communes and production brigades were to establish flood prevention committees or assign cadres for this work. In the city, all bureaus were instructed to establish their own flood prevention offices and assign 3–5 expert cadres to be responsible for flood prevention work. Cadres could also be transferred to them to take care of flood prevention work as a full-time task.[12]

The plan declared that there had to be pre-planned principles of flood prevention for all eventualities. Flood walls in the city proper should not collapse under any circumstances, and areas and periods of water accumulation in city districts had to be diminished. The suburbs also had to be able to overcome disaster conditions caused by a flood. The Hydro Power Bureau and the Construction Bureau together with the River Road Management Units were instructed to inspect all embankments, protective dykes, important sluices, and piers. Danger spots had to be repaired before the flood season began. The most seriously soggy spots in the walls had to be repaired and materials made ready for repairs at other spots. It was also noted that the TJFPHQ had to prepare propaganda materials to fight apathy in order to create a mental basis for victory in the coming struggles.[13]

The plan also indicated how a limited mass mobilization was planned to be carried out in the eventuality of a flood. For this end, the TJFPHQ issued instructions for the organization of city maintenance workers and sanitation workers to act as stand-by flood prevention troops. Further, the PLA troops stationed in Tianjin, the Construction Bureau, Building Management Bureau personnel, and suburban production teams' basic-level cadres and militia were to form emergency teams. These would be used according to the TJFPHQ orders in time of distress. Moreover, city administration basic-level cadres, every *danwei's* militia members and 'robust' peasants in suburban areas were organized into backline (reserve) troops. The word 'robust' indicates a concession to the post-famine realities in the Tianjin countryside. The 'reservists' would be used under special orders in times of flooding. Also flood prevention materials, such as straw bags, galvanized wire, and gasoline were to be purchased and kept ready by the TJFPHQ. All districts had to assign responsible persons for sections of embankments. The principle was to assign a section of river walls in advance to communes and *danweis*, which they would be responsible for taking care of in time of emergency, and arrange inspection patrols according to these regulations.[14] The actual mobilization during the flood did follow these outlines quite closely. Units more or less knew whom to draft into the campaign, where to go in time of emergency and who would assume the command. The city was therefore organizationally prepared for flooding at least in principle.

Also importantly, at the beginning of August just before the flood, the HBFPHQ issued an administrative opinion on how to arrange discharge routes for floodwaters into Gukouwa, which was one of the major depression areas suitable for water retention, lying southwest of the city. It was designated as storage space for floodwaters. The opinion offered complicated schemes concerning the order and method that would be used to divert water into the depression and which flood walls were to be pulled down in the eventuality of flood.[15] In the end the opinion was not carried out as such, but it indicated that *a priori* planning of possible water diversion routes was underway. When the flood hit Tianjin, water diversion was a crucial part of flood prevention work and was conducted under direct guidance of the HBFPHQ.

A report by the TJFPHQ on the eve of the great flood indicates that many of the measures outlined in the plan were also carried out in June–July. The TJFPHQ reported that it had conducted inspections on land and in water of all major embankments, walls, sluices and piers in all of the rivers and reservoirs under its jurisdiction. Suburban and city district industrial units and warehouses had undertaken drainage arrangements and materials had been reserved for flood protection work. The city had organized 15,000-strong standby repair teams.[16] However, the report did also complain that some units were still dragging their feet and demonstrating 'numb thinking' concerning the risk of flooding. Suburban communes and brigades and handicraft units were especially backward in this respect. Therefore, spring repair works were lagging behind schedule and in some of the suburbs drainage channels were only 40 per cent ready.[17] There were also bigger problems. For example the Duliujian drainage channel had only about 60 per cent of its drainage capacity left due to sedimentation and reeds. This was still the situation when the flood hit the city.[18]

On paper, at least, the city was therefore prepared to withstand a big flood, but the magnitude of the 1963 flood was clearly beyond the scope of planners. The idea to mobilize the whole city for a campaign to save it from inundation was not mentioned in the plan and had to be carried out *ad hoc* when the flood hit. However, the organizational principles in the plan formed the basis of the mass mobilization that became a central part of flood prevention work and the TJFPHQ provided a command structure for it in 1963.

The General Flood Situation

From 1–10 August 1963, western and southwestern Hebei Province regions upriver to Tianjin were hit by continuous heavy rain, which created the worst flood ever recorded in Hebei provincial history.[19] The flooded area consisted of the entire Hai River basin. During the severe downpour between 2 and 5 August, on a single day 1,000 mm of rain fell. Such daily amounts equalled or even exceeded average rainfall for an entire year.[20] This overwhelmed many dykes, and flooding hit 101 counties in Hebei Province affecting the lives of some 23 million people.[21] The flood waters were carried to the sea by the Hebei river system's big rivers, the Daqing, Ziya, Nanyun, Yongding and Beiyun, all tributaries to the Hai River (Haihe), which discharges its waters into the Beihai Sea through Tianjin city.

Haihe runs through the central urban areas of Tianjin city and all these rivers merge together in the Tianjin suburbs at different places. Continuous rain and the collapses of important embankments upstream caused two flood peaks around 14 and 20 August during which the metropolis came under serious threat of inundation.[22] At its peak, the flood water level in rivers and reservoirs rose some 4 m above the average elevation of the city proper.[23] If the flood waters had been allowed to flow freely, the main commercial street in Tianjin city centre, Heping Road, would have drowned under 2.1 m of water.[24] It is clear that if such a flood had hit the important industrial city of about 4 million residents the consequences would have been catastrophic.

Controlling the Flood

Soon after the heavy rains had started at the beginning of August in Hebei and Shandong, Tianjin also began to prepare for the flood. At first, the city provided disaster aid in the form of flood-combat material to other Hebei regions,[25] and continued to do so some time after the emergency mobilization to save the city itself had begun,[26] but after 8 August protecting the city itself became priority. By 6 August, some 30,000 people had already been dispatched under local party

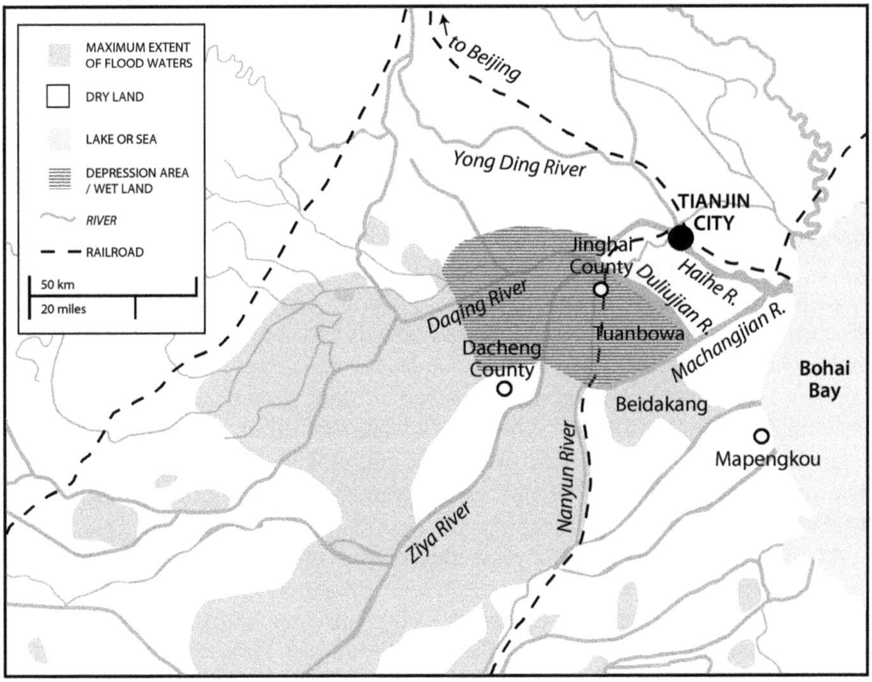

Map 2.1 The extent of the 1963 flood in the vicinity of Tianjin

committee leadership to enforce river embankments. It was reported that by the midday of 8 August they had already raised 200 *li* (*c.* 100 km) of the river banks by one metre. Materials were being transported to the most vulnerable places by truck and boat. Some 9 million bricks and 90,000 straw bags had been made ready for the work.[27] But this was merely a beginning.

On the same day, mobilization shifted to higher gear. Massive flood waters from upstream had begun to reach the Baiyangdian water retention area southwest of Tianjin.[28] The Provincial Party Committee and Government held an emergency joint meeting, making anti-flood work the first priority in the province. During the night of the same day, the same bodies of Tianjin city held their own emergency meeting. Wang Peiren delivered a report to the meeting and estimated that the flood peak would reach Tianjin on 12 August and could easily overcome the capacity of reservoirs and drainage channels protecting Tianjin around the Daqing and Ziya rivers.[29] Based on instructions from the Party Centre, State Council, Provincial Party Committee and Government, the meeting decided that the flood waters would be stopped at the outskirts of the city and then led to the Bohai Gulf. For this, the meeting decided to mobilize 'entire party and people' in the city.[30] Consequently, the following days witnessed a large escalation in mobilization to save the city from the flood.

Four new sub-FPHQs were now established at the Ziya, Duliujian, Beiyun and Jinzhong river sections with the duty of protecting their embankments at any costs. These sections were designated as 'Tianjin's protective screen'. The three first sub-HQs were placed under vice-mayoral command and designated city districts were assigned the duty of supplying them with flood-combat troops. The Hongqiao, Nankai and Western Suburb Districts were to send their people to Ziya River, and the Heping and Hexi Districts were to assist the Western Suburb District in protecting the Duliujian. Jinzhong River was the responsibility of the Eastern Suburb District. In the city, Hai River embankments had to be elevated to 6.5 m along its full length. This duty was assigned to city districts, with 2–3 days to rush to get it done.[31] District people's governments were ordered to establish their own FPHQs and all city administrative units and companies were to establish their own FP Offices, with at least a vice-leader-level director. All major central bureaus of the city had to dispatch members of their personnel to serve in different offices of the TJFPHQ.[32] Army units were mobilized according to existing regimental structures, and their units were under Tianjin Garrison command.[33] The army set up its own FPHQ for frontline areas surrounding Tianjin headed by a lieutenant general-level officer.[34]

Already before the emergency meeting on 8 August, the Hebei Province and Tianjin City Governments had organized a small three-man group above the TJFPHQ to coordinate flood combat between the city, province and other regions. This group, which must have consisted of the Provincial Party Secretary Lin Tie, Tianjin Party Secretary Wan Xiaotang and maybe the Tianjin garrison commander as the third member,[35] decided that the basic policy was to speed up northern water flows (through Daqing River into Haihe) into the sea, while slowing down the southern flows (through Ziya and Nanyun Rivers into Haihe). To this end, southern flood water flows had to be diverted into reservoirs and other retention areas along the river streams before they reached the city.[36] It was further ordered that localities must not take the initiative in protecting themselves. Under no conditions must railroad foundations be dug up or embankments broken up or new ones constructed without authorization. No locality or bureau was to act based on its own narrow sectarian interests and thereby harm the common good, it was stressed.[37] As during the GLF famine when the city had struggled to manage the food production and keep the delivery chain under its unified control, it also sought to control the flood protection work. However, in 1963 the city succeeded notably better than during the GLF.

The main role in water diversion that followed the decision was first played by the three main western depression areas (Dongdian, Wenanwa and Gukouwa) and then the three 'eastern' ones (Tuanbowa, Beidakang and Tangjiawa), which all actually lay to the west and southwest of the city, while the eastern ones lay closer to the sea (see Maps 2.1 and 2.2). The western depressions were filled up in succession during the first flood peak through water diversion measures and unplanned collapses of embankments, then after about 18 August the eastern ones began to fill. Water diversion was not an easy or risk-free tactic even under the best of circumstances. It required pulling down walls in many places and

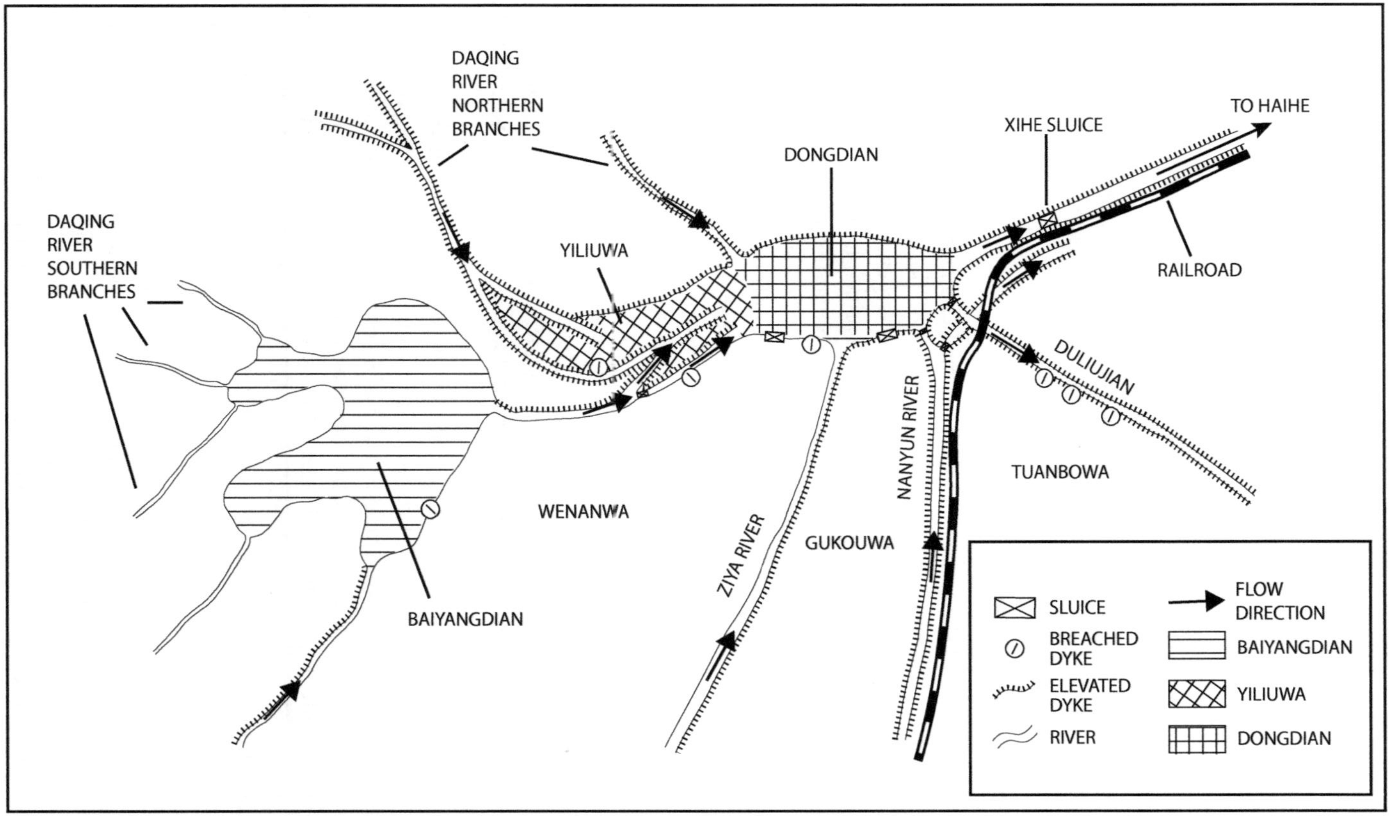

Map 2.2 The key rivers and retention areas in the 1963 flood prevention campaign

practically flooding the Western parts of Tianjin rural suburbs causing a sizable refugee problem. In essence, a larger urban disaster was exchanged for a smaller rural one. The gravity of diversion decisions is shown in the fact that it was the provincial leadership that gave the orders on every occasion when a major diversion was undertaken. After the latest reorganizations of Tianjin city borders (see Chapter 1), many of the inundated areas lay outside of its jurisdiction anyway, so the city could not make such decisions on its own although thousands of Tianjin residents were working on the embankments and drainage channels at these sites. But while water diversion was the key to defeating the flood, it required mass mobilization to succeed as the river walls protecting the city proper had to be reinforced and protected from collapsing under the pressure of the enormous flood waters.

Emergency Response

The TJFPHQ established the rather oddly named Organization Education Office (OEO) to plan the mobilization and deployment of flood-combat troops. Right after the City Party Committee and Government emergency meeting on 8 August, the OEO held its own meeting and decided on the principles of work: the frontline was to be given priority in terms of manpower, but at the same time production in the backline had to be maintained. It was further decided that the flood-combat troops would be organized based on militias and militia structures, and divided into frontline regular dyke protection troops, reserve troops for flexible use, and emergency troops. The dyke protection troops would be mainly drafted from peasants, street residents, administrative cadres, high and middle school staff, and workers from those industrial units where this would not affect production too much. Emergency team workers would be peasants, construction workers and workers from industrial units. The units would self-organize under their district FPHQs, except units designated directly under the TJFPHQ (such as the Construction Bureau and Electricity Bureau personnel). Army units were all to be assigned to act as emergency troops.[38]

When the plan was carried out, 660,000 militia members in *danweis* and people's communes were mobilized into flood-combat units consisting of 20 divisions, 160 regiments, 704 battalions and 3,378 companies. 620,000 people were dispatched to the frontline, and the rest had mainly logistical duties. However, not all were eligible to join the fight. The OEO outlined that the politically suspect 'five elements'[39] were not allowed to participate in the troops. Furthermore, women, children under 16 and the old and infirm were not to be given frontline duties, although in an emergency all could be called upon.[40] The exclusion of the class enemies was based on security considerations of not letting the 'usual suspects' in Maoist society to engage in any mischief on the dykes. The City Party Committee also ordered that counterrevolutionary and bad elements had to be prevented from creating havoc during the struggle.[41] Women and the young, on the other hand, were regarded as physically too weak to take on strenuous job on the dykes. There were even complaints about how a large number of women workers had negative influence on dyke construction.[42]

As intended, this emergency mobilization now touched every *danwei* in the city and its suburbs. It was made known to the city residents in the *Tianjin ribao* on 9 and 10 August, without mention of bursting damns upriver.[43] As planned, the militia played a central organizational role in this first wave of mobilization, when basically all work units with such an organization in the city and its rural areas were ordered to mobilize their militias under the direction of respective party secretaries. These were then at least generally speaking dispatched to the pre-designated sections of dykes, sometimes at very short notice.[44]

The seriousness of the flood also drew the attention of the central government, which approved the basic plans made by the province and the city leaders. On 11 August, the State Council issued an order to Hebei Province instructing it to protect Tianjin city and the Tianjin–Shanghai (Tian–Pu) railroad as well as keeping damaged areas to a minimum.[45] Saving the city was therefore made a national priority. If needed, it was now authorized for flood waters to be diverted to achieve this goal. Vice-Premier Tan Chenlin visited Tianjin to arrange disaster relief with provincial and city leaders.[46] These instructions were also carried out in a few days as the situation took a turn for the worse.

By 12 August, 340,000 people had already been mobilized and 180,000 were working in the frontline where they received the first flood peak.[47] However, after Baiyangdian reservoir filled up and its eastern dyke had to be breached on 11 August, it appears that the city leadership realized that this was not enough. On 12 August, the City Party Committee once more ordered the party organizations in the city to mobilize all party members and the masses to the fight and sent a grandiloquent letter to all party members in the city. The letter declared that it was the party members' duty to see to it that that the mobilization of the whole city took place. Therefore, all members themselves had to get vigorously mobilized. Party members should not fear any hardship or danger, but eat bitterness and to go forward developing the communist spirit of self-sacrifice for others, while being disciplined and following orders. So doing, they would get the masses to join the struggle enthusiastically, it was explained.[48]

On 13 August, the City Party Committee and City Government declared further emergency instructions for the whole city for resisting the flood and mobilizing all party and people to fight it.[49] These were made known to the residents on 14 August in the *Tianjin ribao*. The emergency decree declared that the struggle against the flood required a contribution from every person in the city as well as using all available materials and financial resources for it. All factories, companies, shops, schools, hospitals, streets, etc. units had to join the struggle.[50]

Moving on to the mass mobilization in disaster management was necessitated by the first flood peak, which flowed on 14 August with its full force through the Daqing River and the Ziya Rivers into the Baiyangdian and Dongdian retention areas. Under extremely bad weather conditions between 14 and 16 August, several threatening situations occurred at all sections of embankments. Troops struggled all night long to contain leakages, sometimes with their own bodies. At the Ziya river embankments, 10,000 Nankai and Hebei District men went into the water to protect the walls with their bodies during the night of 14 August in a force-nine gale singing the *Dongfang hong*.[51]

The first flood peak was caused mainly by the Dajing river system that was fed by waters from collapsed reservoirs upstream. By 11 August, flood waters had filled the Baiyangdian depression then a hole was breached to its right-side (eastern) wall to let waters flow into the Wenanwa depression. When this was not enough, another hole was made on 13 August to let the waters flow into the Yiliuwa retention area, which lay to the north of Baiyangdian. However, the Baiyangdian water level kept rising. At the same time, the northern branches of the Daqing river had flooded between 5 and 9 August with waters flowing to the Dongdian depression area, which was the last hold before the city. Because Dongdian was also fed by the Ziya River and discharge waters from Baiyangdian that did not fit in the Yiliuwa retention area, Dongdian began to swell rapidly reaching its flood peak of 8.39 m on 15 August. The situation was described as 'extremely serious' and emergency protective action was taken at the Ziya River and Duliujian (which discharges Dongdian excess waters into the sea).[52]

To control the water situation, the Xihe Sluice, which controls water flows to Haihe from Dongdian, was opened wide. This brought the flood peak into Haihe running through the central city areas, but the Hai River walls held. Luckily, the Beiyun and Zhongdian Rivers that fed the Hai River from the north were not receiving much flood water at this time. Therefore, they could be controlled through sluices and Haihe was able to take in a lot of Dongdian waters, but less than was flowing in. When Dongdian reached the critical water level of 8.00 m, the Duliujian Jinhong flood sluice was also opened to increase Dongdian's drainage capacity, but this did not happen without problems.[53] A section of the Duliujian left wall, the last defence before the city, was 0.5 m lower than the rest of the embankment, and water began to swell over the dyke. Troops and materials were rushed to the spot for emergency repairs, but this was not enough. The provincial and city party committees had to order the Duliujian waters to be diverted to the Tuanbowa depression area and Duliujian's right wall was breached at three points. This saved the situation.[54] After the immediate danger passed from Dongdian, Jinhong Sluice was temporarily closed in order to make repairing Duliujian's walls possible.[55]

The first flood peak was brought under control by these water diversion actions on 15 August. Apart from breaching the Duliujian embankments, the Baiyangdian '1,000 li river wall' was breached once again on 15 August to let more water flow into Wenanwa in order to reduce water flows into Dongdian.[56] The same day, following the HFPHQ plan, Dongdian waters also began to be discharged into Gukouwa and Wenanwa through the Tanli and Guodi flood sluices.[57] These measures made the waters recede at Dongdian and brought the first flood peak under control.[58]

But there was hardly time to take a break. The second flood peak was developing in the southern river section at the Gukouwa depression. The Ziya and Fujin rivers broke their banks and rushed into Gukouwa on 17 August and its water level peaked on 19 and 20 August, reaching 8.94 m. To control this situation, several holes were made in walls to let water flow into Wenanwa and Dongdian. At the same time water flows into Baiyangdian increased once again, peaking around 26 August at 10.5 m. On 24 August, the three reservoirs Gukouwa, Wenanwa and Dongdian had the same water levels at around 8.2–8.3 m and behaved as a united water mass. The waters kept rising until 30 August.[59] The risk scenario was now that if the Gukouwa breached its walls and, together with Dongdian, exerted pressure on the Ziya River's right and Duliujian's left embankments, this would cause serious danger to Tianjin. Another great danger was if the Gukouwa breached the Nanyun River walls and inundated the Tianjin–Shanghai railroad running next to the river. To prevent this, the construction of an outlet river from Gukouwa to the sea was started and walls protecting the city were further consolidated. The construction of protective backline embankments was also stepped up.[60]

To bring the situation under control, the Jinhong Sluice was opened once more and the provincial and city party committees ordered the consolidation of the Duliujian left wall for a distance of over 40 km. For this purpose, the City Party Committee mobilized 63,000 militia members from industrial work units to work on the walls day and night, with 2–3 people per metre.[61] At the same time some 50,000 people were constructing a second embankment behind the Duliujian walls. The other critical section, the Ziya River right wall, had 83,000 men engaged in consolidating the walls.[62]

As noted, the solution to the situation was to reduce water pressure by directing Gukouwa flood waters to the south. Already on 13 August, the TJFPHQ had ordered a flood drainage channel to be constructed between the Gukouwa and Tuanbowa depressions to allow a larger flow of waters from northwest to southeast. The construction was undertaken by over 10,000 Hexi District militia members, PLA soldiers, railroad workers and city construction workers, who worked 24/7 for two weeks.[63] The waters were diverted by blowing up a section of Nanyunhe dykes, which was a difficult task. A team of technicians, old peasants and railroad workers were assigned to the task, finally being able to figure out a way to guarantee the right direction of discharge flows. Holes were blown on 20 August and worked as they were designed. At the same time, troops cleared the earth underneath the railroad, with the help of two earthmovers. In 40 days, the channel drained over half of the excess waters into the depression areas, which was seen as a key to the saving of the city and the railroad.[64]

After the drainage channel was opened on 20 August, Gukouwa waters rushed into Tuanbowa, but because of the limited storage capacity of Tuanbowa, this still threatened the Duliujian left embankment and the railroad.[65] Already on 19 August, the city party and government leaders had ordered a drainage channel to be dug between Machangjian River, whose walls separated Tuanbowa from Beidakang area, and the Beidakang reservoir.[66] On 21 August, air force planes reconnoitred the Tuanbowa situation and, based on the intelligence information, the HBFPHQ and TJFPHQ issued an order to blow openings in the Machangjian River wall.[67] The Machangjian sub-FPHQ was headed by Vice-Mayor Song Jingyi and attended by Vice-Governor Xie. Its troops were mainly from the southern suburb.

The work to build a discharge channel commenced on 21 August and the Machang River walls were blown up on the 23 and 25 August and many more times after that. This stabilized the water situation at Tuanbowa, and water began to recede there on 30 August.[68] Flood waters first went into the Tangjiawa depression area and, after a number of obstacles were cleared away, into Beidakang. However, Beidakang water levels were high already, and it could not take in much more, so waters spread to a number of places before they could run into the sea.[69]

Much of the credit for the success in the flood protection goes to the drainage channels. The last stretch to be constructed was from Beidakang reservoir to the sea through Mapengkou. The channel was reported to have been basically finished on 3 September and was 25 per cent larger than the Hai River.[70] The channel was further enlarged, and additional channels constructed during September.[71] All in all, the drainage channels had more than double the flow capacity when compared to Hai River and formed an essential part of the strategy that led the bulk of waters past the city proper.

At the end of August, the situation was still hazardous with southern suburbs flooded and reservoirs filled to the brim, but heavy rains had ended and drainage channels were being completed. On 2 September water began to recede at Beidakang. While the TJFPHQ still ordered people to be vigilant,[72] the *Tianjin ribao* declared initial victory in the struggle on 7 September.[73]

The first dispatches of workers in strategically important industries had already been sent back to the city on 24 August to resume production. After this, flood prevention troops were gradually disbanded. The policy was to keep the organization and its backbone cadres intact at the dykes until danger had really passed. The HBFPHQ ordered that releasing people from flood protection duties had to proceed according to the water situation and *danwei* type, with military and industry units sent home first.[74] First to go were workers in export and weapons industries as well as in industries supporting disaster areas and agriculture, and those who were needed to 'stabilize city markets'. Teaching staff were also sent home to open schools, along with the 'old and weak' and those with dependents at home.[75]

The departing troops were instructed to clear away the flood protection materials, which they had used, such as communication equipment, tools, reed mats, bamboo poles and sand bags, and hand them in to assigned persons for safekeeping. The HBFPHQ ordered the establishment of a warehouse system to store the material. The disbanded troops were instructed not to take any borrowed tools or equipment with them and keep records of the materials they handed in.[76] The TJFPHQ further instructed that borrowed materials that had been damaged during the struggle were to be compensated to localities.[77]

The city made the return of the flood-combat heroes a great propaganda spectacle. Big celebrations for the troops returned from the frontlines took place in the last week of September, when the most distinguished flood fighters and units were rewarded at a grand gala.[78] As an eloquent ending, the honoured heroes in the PLA were also invited to the National Day reception in Beijing on 1 October, where they were received by Mao Zedong and other party leaders.[79]

Mobilization

The flood provides a prime example of how Maoist disaster governance through campaigns worked. As described above, in order to achieve successful flood control, a massive operation to construct, consolidate and maintain dykes and dig drainage channels had to be organized extremely quickly. This required the engagement of the masses for about six weeks. The emergency mobilization of flood prevention troops, which was decided upon on 8 August and expanded in subsequent orders from the provincial and city leadership expanded mobilization to 961,989 people, who had some 600 km of dykes to take care of. In the end, 371,927 of the mobilized people were assigned to the frontlines, 251,189 to the backlines and the rest had other duties.[80]

Typical of Maoist mobilization, flood protection also relied upon the existing administrative units. The mobilization cycle started with mobilizing groups, which were closer to the bureaucratic core of the party-state organizations. Therefore, professional construction and city maintenance units were the backbone of the early limited mobilization when the City Construction Bureau, the Construction Work Bureau and the Housing Inspection Bureau were ordered to form emergency troops to undertake rush repairs of flood embankments. Other professional units were mobilized for supportive functions, such as the City Health Bureau, which was ordered to transfer cadres to organize medical teams and prepare them to be called out at short notice. Originally, the City Transportation Bureau had to prepare 50–100 cars and teams of drivers capable of driving on the suburban mud roads to be on 24-hour readiness for emergency transportation work. The Telegram Bureau had to take care of communications between the TJFPHQ and the four sub-HQs, where telephone lines had to be constructed and specialist communication personnel had to be dispatched for this purpose.[81]

In the mass mobilization phase people more remote from the bureaucratic core were drafted to the campaign when all city party members and *danweis* were ordered to take part in the struggle.[82] Instructions were now that all *danweis* (including schools, factories and shops) were to send their strongest and politically most aware members to the frontline with responsible cadres to lead them, but at the same times *danweis* had to maintain the backbone of their workforce and leadership to continue their regular production activities.[83] The order therefore indicated how flood mobilization was not to interfere with the city's production duties. Once again, as during the GLF famine, disaster management was a distraction from the city's main tasks. But this time mobilization was ordered from the central government and had its support.

The role of the militias which had been established in the early GLF was crucial to mobilization. As the TJFPHQ OEO report on the mobilization work noted, without the militias in work units, establishing such disciplined organization in such a short time would have been 'very difficult' and added: 'It seems that [the] "everybody is a soldier" [principle] is not only needed to [oppose/defend against] class enemies and imperialists, but it is needed also in the struggle against nature'.[84] The leaders of militia troops were usually leading cadres in work units and all flood-combat organizations established their own party cells taking charge of leadership and propaganda work in them, thus replicating the party-state organizational principles in their formation. In mixed units, i.e. units with members from different *danweis*, the leading cadre was dispatched by the TJFPHQ or local FPHQ. Well-trained militias could get organized quickly. A Heping District cultural system 1,000-strong militia was able to get assembled in two hours at a night drill on 10 August. That time they were still sent back to sleep.[85] However, it is unclear how many of the city's militias were so well-trained. For example, at the Ziya River flood prevention site, militias were subjected to discipline education and told to learn about discipline from the army.[86] Nevertheless, the militias' key role in the mobilization is indisputable.

The Ziya River sector also gives a good example of how the troops were organized by sub-FPHQs. Troop organization followed military organization with teams, platoons, companies, battalions and regiments. Work units guaranteed at least one militia platoon (30 men) for the task, but in reality a 'platoon' could be as small as five men. City districts were assigned their own sections on the Ziya embankments, which were then further divided into regiments, battalions and company sections. PLA troops acted as emergency teams under the Ziya FPHQ, but had their own command structure. Militia battalions and companies had their party branches and Youth League organizations. On every level, most of the questions were decided by party committees.[87] Flood prevention troops took the party-state to the dykes with them.

Apart from the administrative division based on *danweis*, there was a functional division of labour as well. As noted, the flood prevention troops were divided into frontline and backline units. Ordinary mud workers were also different from the emergency teams. The latter had the strongest members in them and their mission was to rush to dangerous spots to protect the walls from collapsing and help to block leakages with their own bodies if needed.[88] All army units taking orders from FPHQs were assigned such shock troop tasks,[89] but civilian *danwei* units also had their own internal emergency teams set up under FPHQ instructions. Some *danweis* selected the best swimmers for special 'fighting teams' to do underwater and rescue missions. Some *danweis* also drilled their emergency teams for action during lull periods. *Danweis* also organized their own first aid and transport teams.[90]

Table 2.1 The composition of flood prevention troops in 1963

	1st line	*2nd line*	*Total*
Workers (SOE)	198 409	91 220	289 629
Collective industrial workers and street residents	84 410	67 265	151 675
Commercial sector workers	41 152	70 006	111 158
Peasants	100 211	89 732	189 943
Soldiers	53 700	—	53 700

Source: FXZL 15.10.1963, 197; numbers exclude those participating in logistical duties

The breakdown of the composition of flood-combat units in Table 2.1 shows that the mobilization was predominately urban and civilian in its nature. Workers were the largest group of drafted people, about 72 per cent of them serving in the frontlines and the rest in logistical duties. The 'political' composition of the troops showed that 81,825 of troops were party members, 49,425 of whom served 'on the dykes', or about 60 per cent of total drafted party members. 85,134 were Youth League members, 48,882 of whom served on the dykes.[91] Party-state presence was therefore strong in the troops.

The TJFPHQ instructed sub-HQs to set up quality inspection teams under the formula of 'joint three' of cadres, representatives of the masses and experts in inspection teams to constantly patrol the dykes.[92] The teams were a response to the justified concerns about the quality of rush construction work, and they were reported to have notably improved the quality of newly constructed walls.[93] The role of construction experts in providing technical instruction, supervision and inspection functions was also emphasized repeatedly. This stress on experts was probably one of the most un-Maoist features of the whole campaign, but in 1963 it was clearly possible.[94] When victory was declared in the struggle at the end of September, a column in the *Tianjin ribao* noted how it had been made possible by using and trusting in science and scientific personnel as well as uniting them with the people.[95] This echoed the 'walking on two legs' thinking of the time. Such an approach was clearly influenced by the Maoist policy of mass science and technology, but also indicates the return to more rational policies after the follies of the GLF.

Most of the technology used in combating the flood was rather rudimentary, however. Militias were instructed to take their own spades, baskets, etc. with them when mobilized.[96] The standard was to bring a shovel and two baskets for five men to carry earth or one large basket carried by two people.[97] Militias were also told to take along camping gear, washing tools and utensils, supplementary foodstuff, pans and pots, fuel, etc. The TJFPHQ also ordered the troops to be supplied with warm food at least once in two days.[98] When arriving at their designated segments of embankments, they put up their red flags on the walls and began working with what they had.[99] At the beginning, many units were dispatched to the dykes with only some dry food and simple tools with them with little shelter from mud, mosquitoes and the rain that fell continuously in the first weeks of the flood. Many people simply slept on the dykes. Their supply of rations was adversely affected by the abnormal weather, and this had a major influence on the health situation.[100] The Liangwangzhuang FPHQ report described the early days at the dykes as 'not eating warm food, not drinking boiled water'. It also described the situation as 'muddy roads and not enough tools, at night no lights'.[101]

When the troops' stay became prolonged, more attention was paid to housing and their other needs. *Danweis* were instructed to manage their own troops' living conditions at the frontline and pay closer attention to shelter, food and drink. Most *danweis* established Livelihood Management Offices for this task.[102] It was reported that many *danweis* also set up clinics, and brought home-cooked food and water to the workers on dykes. They also delivered anti-mosquito pesticides. Many *danweis* divided their troops into two shifts so that people could get enough rest.[103] Many supply depots and service points were established along the frontline by the City Commerce Bureau where for example bicycle, shoe and clothing repairs were made.[104] Propaganda also claimed that the Commercial Bureau set up stalls to sell fruits, cakes and melons to the troops. There would also be barbers and broadcast stations, and newspapers would come fresh to the frontline.[105]

Various methods were used for temporary housing of the troops. In some communes, peasants were reported to have offered to house the troops and peasant stoves were also used for cooking for them.[106] Other units erected their own shacks to prepare food and offer shelter.[107] Other units provided canvas and/or straw tents for their members.[108] In flood propaganda, military units were praised for their tenacity in spending their nights outside, not bothering the peasants with their needs,[109] which tells about the housing difficulties and how the propaganda tried to turn necessity into virtue.

Problems did also arise in the organizational work. As the reports admitted, because of the urgency of the struggle, some organizational work had been haphazard, ideological education had not been strong enough, and the people's understanding of the water situation had not been good enough. Too many cadres and members of the masses let down their guard, especially after the first flood peak had passed. Quality problems with many of the elevated river walls were also admitted. Transportation had problems, needed materials did not always arrive or went to the wrong places and some of the materials were stolen. The typical Maoist solution was to call for better ideological education of the masses. Better organization was also called for, as was further consolidation of the walls, as well as better inspection and emergency work.[110] In September, it was reported that the organization and general order of the struggle had improved compared to their haphazard nature in the beginning.[111]

But how did the party-state motivate people to actually participate in the campaign? While joining it could be just ordered from above, creating genuine enthusiasm for it was another matter. Basically, the party-state could use either material or moral incentives. In the campaign against the flood, material incentives were related to grain rations and salaries. Draftees' supply standards were actually not that bad. Frontline workers' rations were set at 45 *jin* of grain per month and even suburban commune members were provided an extra 0.5 *jin* a day.[112] Selected products were also sold to frontline workers without coupons, such as soap, toothpaste and rubber boots – although in limited numbers. The central government sent 600 boxes of cigarettes and 5,000 tons of fuel to the frontline to support and console the troops there; it also provided some of the extra grain needed for flood work subsidies.[113] State and public sector workers in the frontline continued receiving their salaries as usual. Those paid by piece were paid salaries based on their production the month before the flood.[114] Without such material incentives, popular participation in the campaign would probably have been far less enthusiastic.

The Propaganda Campaign

However, apart from such material incentives, the Maoist urban state sought to overcome the problems in mobilization through the standard means of moral persuasion and ideological education through propaganda. In general, propaganda was needed first to mobilize peasants, workers and members of other urban strata to move to the frontline and, once there, keep them in their posts long enough to do their jobs properly. A large propaganda effort to keep people motivated and let them know what was expected of them was an essential part of the whole disaster management process in 1963 with the official organ, the *Tianjin ribao*, as its main channel. It provided daily torrents of flood-related news to the participants in the struggle. Indeed, at the peak of the mobilization there was hardly any other news in the paper than that related to the flood. The common method was to offer examples and models for readers to emulate with the aim of teaching people to look at the mobilization through the lens of class and see themselves as representatives of their respective social backgrounds.

By doing this, propaganda turned the whole episode into one big moral narrative on cultivating revolutionary morals and class comradeship. The great battle against nature was turned into a moulding experience in socialist ethos with its emphasis on loyalty to the Party, self-sacrifice for the common good, fraternal class love, collectivism, discipline, and arduous and unyielding struggle to the very end. Such transformation of individuals through practice was highly Maoist. As such, a flood came only next to a war as an opportunity for such a great education effort for the Party.

Propaganda was already mentioned in the Tianjin City Flood Protection Work Plan, which stated that flood-combat propaganda work had to be strengthened to get rid of the 'numb thinking' of city residents. 'Proper thinking' was to be the foundation of victory in flood-combat it was announced.[115] When the flood hit Tianjin, propaganda work also shifted into a high gear. The city authorities instructed that ideological education had to be carried out on the embankments and combined with the ongoing socialist education campaign.[116]

All in all, while the physical mobilization and anti-flood struggle left people with myriad impressions and experiences, it was the propaganda machinery's duty to tell people what to think about it in a way that motivated them to participate, and at the same time to tell a politically correct story about the regime and its involvement in the crisis. After the dismal failure of the Great Leap Forward, this time mass mobilization under its leadership proved to be capable of delivering large-scale success. Not so surprisingly, the Party sought to extract full political dividend from the events. In 1964, an edited volume was published to celebrate the victory. In it the editors were not shy to make the point of the Party's role in the whole effort in the introduction by stating that: 'This victory in anti-flood struggle was a victory for the Party's leadership.'[117] Under its, and only its, leadership, the disaster had been turned into a great victory, or as the writers put it: 'This exceptionally large flood did cause us some damage, but when we struggled against it to victory, we received many valuable lessons and beneficial experiences. This made a bad thing into a good one.'[118]

During the emergency mobilization, the Party preferred to increase the legitimacy of its rule by comparing its efforts to the failures of the previous regimes. In a typical article describing such comparison, an old soldier,[119] worker[120] or peasant was made to recall the events in the great flood of 1939, when Tianjin had suffered greatly from inundation. Here a contrast could be drawn with all aspects of the mobilization in 1963: the role of cadres, sister regions, all classes, soldiers, the national and local leadership, and the relief effort. Remembering such a bitter past was said to give people strength and motivation in the struggle under the New Society. For example, in one such article an old peasant was recorded as having argued emotionally for the present regime in the following way:

> In the old society, the masters showed no interest in [relief work], but could the present people's government increase its concern for us and the struggle [any further]? You young people have not suffered in your lives! When there was a big flood back then [1939], it destroyed crops and houses. There were three of us, and we had nothing to eat. We had to walk to Dongbei to do hard manual work. My old father perished. Now the world is really not like that! With the help of the Party and government help and leadership, we can get through all hardships.[121]

Notable is also what was missing in propaganda: Chairman Mao Zedong and his thought were mentioned only cursorily. This also happened only at the end of the campaign to save the city, when it was celebrating victory and the Party elaborated the reasons of the triumph. Here, Mao Zedong thought and the Chairman's leadership were given due credit, but otherwise neither he nor his thoughts were mentioned in the propaganda during the mobilization. The struggle was conducted under the leadership of the Party, not Mao, which was probably related to the failure of the Great Leap Forward and the weakened position of the Chairman after it. Paradoxically, it was in this struggle where the Maoist mass line actually showed its strength.

Propaganda designated Tianjin workers as vanguards in the fight against the flood. Other parts of the population were instructed to 'listen to the voices of the workers and follow the working class' heroic example', because the working class was claimed to possess the correct selfless and unyielding attitude required in the struggle.[122] Examples of politically correct acts of working-class members in the frontline were elaborated in the *Tianjin ribao* for everybody to emulate and learn from. For example, during the mobilization against the first flood peak, the following story was told: Hongqiao and Nankai District *danweis* had dispatched over 20,000 people to work day and night on the dykes where they were working hard carrying heavy baskets of earth. Using pushcarts in this work was a real help. Therefore, everyone was upset when a pushcart suffered a puncture. Luckily, a worker named Sun Shouwu immediately cut a piece from his own bicycle tyre to repair the pushcart saying: 'My small loss is nothing compared to the security of the whole city, I should serve the public.' As a consequence the construction task did not suffer a setback, and was accomplished in time, it was reported.[123]

It was further told how when dispatched units of workers arrived at their assigned sites, they discovered that there were more people there than there were originally supposed to be – eager workers who had been ordered to stay behind had actually sneaked into the ranks of the flood-combat troops.[124] Another story that was told was how men whose wives had just given birth to a baby would go to the frontline and not stay behind, even if it was their right to do so.[125] Such stories were obviously written to imbue the campaign participants with correct self-sacrificing ethos. It is difficult to say anything concrete about their veracity, however. The tasks at the dykes were strenuous at best and dangerous at worst; as propaganda itself put it, the work was tough. It meant endless moving of earth and materials with blistering and bleeding feet, hands and shoulders, while getting little or no sleep because of the lack of shelter from swarming mosquitoes, heavy rain and wind, at least before the troops were able to erect proper shelters for themselves.[126] In addition, danger was ever-present. People could easily slip, or get injured in multiple ways common to any large construction site, and could also drown. Many worried about how their families were doing back in the city. Staying behind, if possible, must have been a tempting option for individuals so propaganda had to be written accordingly.

The Party used the opportunity to mend relations between it and the people it led, relations that the GLF had greatly strained. The unity of the party cadres with the masses they led was emphasized time and again in propaganda. In a story 'Bravery at the Ziya River' in the *Tianjin ribao*, Party Secretary Wang from a Tianjin electronics factory was described to have taken the lead in preventing a threatening leak in a dyke by going out into the rain to rally the people to reinforce the dyke, shouting: 'Party members, youth league members! We have to resolutely protect the dyke, never let the flood waters come through…'. As the report continued, he had not even finished speaking when the whole body of workers rushed to the embankment saying: 'We are not Communist Party members, but we belong to the flood-combat army. We want to block the water together with the party and youth league members!'[127] Similar language was used in terms of the relationship between soldiers and the rest of the flood-combat troops.[128]

One other major theme of the propaganda effort was, on the one hand, to mend relations between party cadres and the common people and, on the other, between urbanites and peasants. Worker–peasant unity was especially emphasized,[129] and getting peasants to care for the common good and goals was seen as a specific problem. Therefore, examples of peasants leaving behind their own communes and villages and coming to dykes many *li* away were praised in the *Tianjin ribao*. In articles, they were described as doing it so eagerly that even after travelling long distances they did not want to take a rest, but rushed to the embankments to work saying such politically correct things as: 'The flood is great, but not as great as our determination to rescue Tianjin.'[130] Or: 'Tianjin is ours. Tianjin is a cornerstone of our socialism. We have to rely on her in order to equip our agriculture. In our mechanization and electrification, [the city] provides us with the materials to conquer natural disasters. We really have to protect Tianjin.'[131] Such articles probably indicated friction between urban and rural residents working side-by-side on the dykes after the GLF famine. For peasants, the mobilization meant that they were made to serve the city at risk to their own lives and at the expense of their own crops once again, as in many cases their villages and fields were inundated.

Ideological work was especially stressed when there was nothing serious going on. Therefore, reporting on the urgency of ideological work increased from 18–22 August and then again gradually from 1 September onwards. This coincides with the lull period between the two flood peaks and the end phase of the flood. This suggests that keeping the flood-combat army at its posts when the danger did not seem imminent was regarded as a problem by party leaders. Both workers and peasants questioned the necessity of staying on the frontline and were concerned about how their families were doing while they were away for such a long period of time.[132] The cadres were instructed to constantly fight against such 'slack' vigilance. They for example convened meetings to explain the water situation and conveyed the city leadership's instructions and encouragement to the workers. Other ways of keeping motivation up was by organizing work competitions[133] and giving awards for the best performances.[134]

A specific type of leading cadre propaganda activity was the morale-boosting visits to the frontline by high-ranking leaders, which included many provincial and city leaders. During these inspection visits, the cadres showed their care for the conditions of flood fighters on the frontline. Reportedly, the instructions they gave provided the people with lots of encouragement,[135] and they were anyhow prime photo opportunities for the leaders. Cultural troupes also toured the embankments to comfort and entertain the frontline troops.[136] They even composed pieces of music and poetry for the fight against the flood.[137] Propaganda work was therefore not all about serious ideological education but, as during the GLF famine, propaganda officials also tried to offer light entertainment as a way to keep people's spirits high.

Keep Production Going!

Apart from the socio-political structures that made quick mass mobilization possible, the command economy also played a key part in the campaign. Repairing and consolidating dykes and digging discharge channels required huge amounts of various raw materials and commodities such as sand bags, reed mats (to protect the surface earth of the dykes from waves and wind), rope, wire, lanterns, rocks and various tools. Moreover, the troops had to be supplied with food, clothing, drinking water, medical supplies, fuel and shelter. As in any other war, logistics were the key to success. To this end, the Tianjin Party Committee ordered all companies and *danweis* in the city to support the struggle under the slogan: 'Need people you get people, need cars you get cars, need electricity you get electricity, all for supporting the fight against the flood.' In this vein, on 8 August the TJFPHQ ordered all city bureaus that dealt with relevant materials to prepare to supply anti-flood materials when needed.[138]

It was further ordered that the frontline had priority in getting the materials it needed. The City Commerce Bureau was to supply the frontline in cooperation with *danweis*, whose responsibility it was also to provide the members of their own units with food and other necessities.[139] Banks issued loans to companies for the production of flood-combat materials, but also to suburban communes and production teams to meet flood-combat expenses. For this purpose, it was ordered that all credit cooperatives had to have personnel to handle the loan requests day and night.[140]

The TJFPHQ had the authority to issue orders to the city's companies on needed materials. These in turn could also rely on other *danweis* in suburbs or outside Tianjin to fulfill orders, if their own reserves were not enough.[141] The City Commercial Bureau and units under it played a crucial role in this work. The bureau organized the distribution of necessities to the frontline, but also other material. For example, the City Local Groceries Company had four cadres assigned to the TJFPHQ constantly waiting for orders. When something was needed, such as bamboo poles, they relayed the shipping list orders to the appropriate *danweis*. Many companies were reported to have also begun working during the night to meet supply duties.[142] Companies had responsible persons on duty 24 hours a day, taking care of the delivery of required products.[143]

Other bureaus also participated in logistics. The City Transport Bureau assigned cars, boats and trains to transport material. If a dispatch bill had the character for flood, *xun*, in it, the product had priority.[144] Some 860 km of telephone lines were drawn and wireless communication technology was also distributed to the frontline by the City Telephone Bureau. The City Electricity Bureau arranged power lines to some parts of the frontline.[145] The Health Bureau sent medical personnel to the frontlines. They were instructed to engage in sanitation and health work in advance, investigating for contagious diseases, providing medical supplies, paying close attention to food hygiene, and seeing to it that field latrines were constructed.[146]

It was further ordered that the home front had to take care of production work and continue the ongoing campaign to develop and increase production economically. The slogan, reminiscent of the GLF, was: 'Less people does not mean less production'.[147] Indeed, especially important industries such as those supporting agriculture, disaster areas, the army and exports had to strive to surpass their state plan quotas. At the same time, markets had to be kept stable and speculating and profiteering had to be eliminated.[148]

These instructions were demanding since the factories had sent a large proportion of their best workforce to the frontlines, usually the most able-bodied men. It was also these industries that got their men back first during demobilization of the flood-combat troops in September. Reportedly, many factories sent 40–50 per cent of their staff to the frontline,[149] but the *Tianjin ribao* reported 'victory messages' from factories claiming that they had even increased their production under the influence of the example of the selfless struggle of their fellow workers on the frontline.[150] It was claimed that some factories had been able to produce over their quotas despite shortage of manpower by being able to use equipment more effectively and by making those left at home work overtime voluntarily. The left-behind workers and cadres were sleeping and eating in factories to make this happen.[151]

Some production articles were also dedicated to agricultural production, where the basic message was the same. For example, in mid-August it was reported that all of the suburban collectives had fulfilled their autumn vegetable-sowing plans for the year. As the communes had responded to the City Government's call and sent their workforce to the frontline as soon as they had been ordered, this was a good result from the women and the old who were left behind to do the job.[152] The still-lingering influence of the GLF production propaganda was evident in these reports.

Mobilization was predominantly local in its character, but assistance from other parts of the country also had a notable impact on the struggle, and its propaganda value was utilized to the full. It was reported that a total of 18 provinces and cities sent help to Tianjin.[153] Especially the Northwestern region, Beijing, Shanghai and Shandong supported Tianjin through materials such as sandbags, grass mats, hemp rope, stones, torch lights, electric wire, mosquito repellent, medical supplies, bamboo poles, etc. Shandong Province even diverted some of the flood waters to its Enxianwa retention area.[154] Beijing sent Tianjin cars, communication equipment, and experts such as scuba divers and telegraphers.[155] Propaganda claimed that sister regions had responded to Tianjin's plight with the spirit of 'Whatever Tianjin demands, we will support it; however much it demands, we will provide; whenever it wants it, we will deliver!'[156]

Aid from other regions began to flow in already on 13 August.[157] This was made possible by orders from the Party Centre. Also the regular lines of command were followed in organizing help but, as during the GLF famine, during the flood mobilization Tianjin seems to have had a direct connection to central government, by-passing the provincial level. Therefore the TJFPHQ could operate through the central government when it requested material from outside. In practice this happened through the National Main Office for Supply and Marketing Cooperatives, which issued an order that all cooperatives in the nation had to 'honestly' respond to flood-combat demands from Tianjin within 24 hours. It was reported that orders and positive responses to them could be made within half an hour, as in the case of requesting grass mats from Liaoning which the TJFPHQ issued to Beijing Main Office for Supply and Marketing Cooperatives, which in turn relayed it to Liaoning. The Ministry of Railways made flood transports a national priority.[158] Disaster control could therefore also benefit from the larger system of command economy, but had to follow the hierarchy and rely on the central government agencies to issue its requests to other regions.

City *danweis* also acted on their own in the acquisition of material, and dispatched their own purchasing personnel outside Tianjin to visit companies that were producing the required goods. Reports indicate such activity at least in Jinan[159] and in Hangzhou, where the local rope factory opened its warehouse to purchasing personnel dispatched from Tianjin to acquire supplies.[160] This indicates that companies' own purchasing personnel who had become active during the GLF, and caused a headache in Tianjin, were now turned into assets working for city disaster management.

Disaster Relief Efforts

The 1963 flood never became a disaster in Tianjin city proper. Nevertheless, it did wreak havoc in the city's suburbs and also a number of urbanites did fall victim to it through death and injury on the dykes. The largest group of disaster victims was the peasant refugees, who had to evacuate their villages when waters were diverted to retention areas. After the flood prevention phase, the disaster management cycle turned to relief effort, reconstruction and to the question of compensation for the victims.

First there was the question of the injured and those who had died in the campaign to save the city. After the flood began to recede, units were instructed to inspect and evaluate the victims for possible martyrdom. The flood had been devastating in Hebei. It had affected the lives of 23 million people, killing 5,119 and injuring over 430,000 people.[161] Officially, however, only 65 people died as 'martyrs' in the flood prevention campaign in the whole province. Twenty-nine of them were soldiers (five officers or underofficers, and one soldier from the airforce). Nine of the dead were from Tianjin city, seven from the Eastern Suburb, one Eastern Suburb commune-level secretary, four collective team or vice-team leaders, two ordinary peasants, and two workers.[162] These victims were declared revolutionary martyrs, which meant that their families were given extra benefits from the state. According to provincial instructions, a ceremony was to be held for each martyr where the exemplary nature of the deceased was to be emphasized. At the same time, their families had to be prompted to turn their grief into energy to be used in production and to commemorate their dead through labour.[163]

As for the injured, provincial instructions and legislation laid down a general framework of compensation, which followed the pecking order in Maoist society in general. For the workers in public employment, compensation was to be paid according to 'Interim Regulations on Compensation for Revolutionary Workers' Injuries or Fatalities'. The families of the deceased were to be paid the burial expenses and a one-off compensation following the nomenclature ranking of the person when still alive: a grade 13 or higher cadre's family would receive 600 *yuan*, while a family of a grade 25 cadre would receive only 180 *yuan*. The injured in public employment were to be paid their full salary during their recuperation time. In case of permanent disability, compensation was to be paid according to the degree of disability and could be as high as 62 *yuan* a year. For workers in companies, the instructions were to rely on labour insurance and, if a company did not have any, then they had to follow their own regulations. If they had no regulations, then the management of the company was told to follow regulations in companies in the same sector.[164]

The rest of the grassroots cadres and people (*min-gong*) drafted to the campaign were to be compensated according to the 1959 Provincial Regulations on Compensation for Injury or Death in Participating in National Economic Construction. This included paying burial expenses and a 200 *yuan* compensation to the family. The injured would get their recuperation medical costs paid for by the drafting *danwei*. Disabilities were to be compensated according to the severity of injury with total disability fetching 400 *yuan*. Disabled peasants were to be taken care of by their commune and work team following these regulations. The state 'could' help in providing assistance if they were short of funding.[165]

Not everything was compensated. As the city instructions in October noted, crops and buildings ruined in the operation were not compensated. As far as the damage had affected grain production, this had to be taken into consideration in procurement work. Instead of compensation, all districts were told to propagandize the great significance of the struggle, and merge it with the ongoing socialist education campaign explaining how it showed the correct relationship between an individual and the state. People were to be told how only 'a small amount' of crops had been destroyed, but the 'big village' [i.e. Tianjin], the lives of the people and property had been protected. As for buildings, mud huts (*tufang*) were compensated 80–110 *yuan* and brick houses 150–200 *yuan*, each depending on the individual cases. Materials from demolished buildings were to be returned to the owners, but this might have proven quite impractical in reality. Damage to public *danwei* property was not compensated at all.[166]

Handling the Refugee Problem

In the end, flood control was successful and Tianjin was saved from inundation because of the water diversion. However, this created a refugee problem that had to be taken care of, and later the city had to face the challenge of rebuilding. While mass mobilization had been the main method of flood control, these later phases of the disaster management cycle witnessed more limited use of campaigns in a more supplementary role, while city bureaus took over managing the lives of the refugees. They had to solve the challenges of feeding, sheltering and finally repatriating refugees.

The depression areas used for storing flood waters and slowing their flows were normally largely used as fields, but in 1963 water also inundated peasant dwellings and entire villages. The chosen strategy of water diversion was therefore in effect a decision to let large parts of southwestern rural suburbs drown. In Jinghai, 132 of 342 villages were totally inundated by the flood and 163 suffered from it. In Dakang District, 56 of 59 villages suffered. It was ordered that the peasants from these areas be evacuated and 3,000 cadres with cars and boats were organized for the work. In Jinghai, some 190,000 and in Beidakang District 40,000 people were evacuated. Almost 20,000 people were rescued from waterlocked villages by boat with the active help of the army,[167] which shows that in some cases information about leakages did not reach villages in time. According to instructions from provincial authorities, the order of rescue was: people first, then cattle, foodstuffs and last property.[168] Evacuation was reported to have been successful, and in Jinghai County a county gazetteer would note 'not a single person died'.[169]

Hebei Province convened a special meeting for instructing disaster relief work at the end of August and early September. At the meeting, promoting production and disaster-relief work were declared the most important tasks of the province.

It was declared that the situation had to be dealt with according to the Party's and people's 'great tradition' of when 'one region suffers from a disaster, all other regions come to its aid'. It was also ordered that regions in the province that had not met with disaster had to mobilize their masses to increase production and food production in order to help disaster regions. Therefore, production quotas had to be more than fulfilled.[170]

Even before these instructions, Tianjin's authorities had reacted to the refugee problem. The official tone was totally different from the GLF famine. City leaders admitted a direct connection between the decision to flood the southern suburban districts and the refugees' plight. When at the end of August the City Government issued instructions on handling the refugee problem, it began by noting how diverting flood waters had been one method of protecting the city and peasants in areas into which water had been diverted had made great sacrifices in their property, living quarters and crops. Such great Communist spirit in following the interests of the nation and sacrificing oneself for the others was really worth respect and esteem, it was declared, although in reality the peasants hardly had had any say in the decision. Therefore, the city residents had to now show high political awareness and make arrangements 'warmheartedly and appropriately' for feeding and accommodating refugees and providing them with health services while they stayed in Tianjin.[171]

Accordingly, the City Party Committee and Government issued an order to: 'Receive warmly and take care appropriately; all who come [as refugees] are welcome. All [refugees'] problems must be solved.'[172] To handle the refugee problem, the City Government organized the Refugee Relief Committee (*Shourong jiuzai renmin weiyuanhui*) that was divided into the Arrangement Group (organized and run by the City Bureau of Internal Affairs) and the Supply Group.[173] City and district leading cadres were personally put in charge of relief work. Headed by a City Vice-Mayor, Lou Ningxian, city finance and other related bureaus also assigned cadres to take part in its work. As during the GLF famine, all city districts were ordered to establish their own *shourong* offices and reception points for refugees. Unlike before, however, these offices were this time actually helping the refugees.[174]

However, contrary to official rhetoric, refugees were not always (if ever?) treated like heroes. During the flood prevention phase, mobilization propaganda had been used to emphasize urban–rural unity and the general policy line towards rural refugees seemed generous enough. Nevertheless, the actually adopted policies show how even during the crisis the urban state actually sought to keep intact the Maoist political urban/rural division as far as possible, as it had done throughout the GLF famine. This artificial political division was softened a little because of the extraordinary circumstances of a major natural disaster and the man-made component of it, but remained basically in force, as the city was keen to control the flow of peasant refugees into the Tianjin urban area.

At the beginning, the city was ill-prepared to house refugees. When the first victims of the flood began to arrive in August, first from around Jinghai County and then from Beidakang, some of them ended up living in the streets. For example, the

City Disaster Relief Committee Bulletin reported that 19 refugees were living in the alley crossings and the food markets in Nankai District. Some of the refugees had no way to make a living, and had started begging. For example, a man named Liu Jinding and his family members had come to Tianjin to seek help from his nephew, but his nephew had difficulties raising his own family, so Liu Jinding's wife had to go begging carrying their three children with her. Some of the refugees had found other ways to make a living and engaged in illegal activities, such as the black market or theft. The worry was that the situation was harming city safety.[175] When housing was found for refugees, it tended to be insufficient and crowded.[176]

A distinct geographic strategy was adopted to solve the problem of housing the refugees. The bulk of the responsibility for this was assigned to the suburban production teams and households. Furthermore, *shourong* stations were ordered to be located specifically in the border areas of the districts. This was justified as making repatriation easier and in order to protect public order in the city. Refugees could enter the city freely, but had to be told that they could not roam in the city at will. It was specifically ordered that refugees from other regions who entered the city should be prevented from becoming hooligans, living on the streets or building their own huts. However, it was decided that they too should be provided the same amounts of relief aid as the refugees from Tianjin counties. In the end, their very low numbers probably helped in making this decision.[177] The fact that so few refugees made it to Tianjin from other parts of the Hebei Province might have been related to the active policy by Hebei, acting under central instructions, to prevent outflows of flood refugees into cities and other provinces.[178]

To reduce the inflow of refugees into the city proper, at the end of August the city authorities further ordered that refugees who could not be repatriated relatively quickly should be sent to the *shourong* stations in the suburbs.[179] Under this policy, about four-fifths of the refugees were settled in the suburbs. However, while city policy was clearly designed to keep as many refugees away from the city proper as possible, it was at the same time relying on the family ties to provide for the refugees when possible. Therefore, some of the refugees housed by their relatives did gain entry to the city proper. By the end of December the nearby suburbs had accommodated 173,707 refugees, while 30,122 people had been accommodated in the central city districts by relatives or friends.[180]

In the suburbs, the *Shourong* Committee was to decide how many refugees could be sent to each station. However, s*hourong* stations were not refugee camps. It was instructed that the size of the stations had to be kept small and economy had to be kept in mind when they were established. In the city proper, street offices and police stations were made responsible for street-level *shourong* and propaganda work for the refugees. In the suburbs, this was the responsibility of communes,[181] but the actual task of finding accommodation for the refugees was the responsibility of production teams. Refugees were to be distributed based on available room in peasant households. In principle, one family should have had one room, but in case of too many people, less space was allowed. At the same time, it was instructed that peasants must not be made to sleep in the open.[182]

Production teams (villages) were therefore mobilized as a main implementer of the refugee housing policy. According to reports, at the end of November in the West, South and Northern city districts refugees were given on average 1.2 rooms per household. For buildings (outhouses, barns, etc.) that were missing a *kang* (a heated bed) 'the masses' had built 6,292 of them, and prepared doors and windows in 4,805 cases.[183] According to official standards, refugee temporary houses had to have a cooking possibility, a *kang* for warmth and security to survive winter. To make this happen, local cadres in all districts were instructed to arrange constant consultation on and inspections of refugees' housing situations. If inspections revealed problems, rush repairs were ordered to be carried out. However, as late as December the City Government reminded local cadres that refugees' housing problems still had to be solved properly.[184]

Feeding and Clothing the Refugees

Food, clothing and providing medical care for refugees posed another problem that was partly solved through the routine activities of professional organizations, partly through mobilization of the population for donation drives and partly through mobilizing the refugees for self-help. Special urgency in this matter was added by the fact that the national economy was just recovering from the Great Leap Forward. Tianjin had just repatriated roughly the same amount of people to the villages as part of GLF famine management that had now returned to the city (see Chapter 1). Now the city had to once more find food for extra people. The problem of clothing was also important as the winter drew near.

The question of how to feed the refugees was especially important in the post-GLF famine situation. The grain situation in the city was far from abundant and with all its welcoming rhetoric, the city maintained notable frugality in providing disaster relief food to refugees. To begin with, it was instructed that those refugees who brought their own food and grain with them would not be fed temporarily. Those who did not have their own food with them were provided with grain to eat and funding for living expenses, but only under strict rules and procedures.[185] During the emergency mobilization phase in August–October, the City Government provided food, firewood and other necessities to all refugees for free. From November, however, relief aid was provided only to the neediest households through the method of 'appraisal meetings' (*pingyihui*). In principle this was a mass line means-test method used in Maoist social policies in general,[186] and it indicated how also the political divisions of class background were reinforced under emergency conditions. Indeed, at the end of November the City Government issued instructions where it was clearly emphasized that the class line had to be carried out in relief work. To this end, the poor, low and middle peasant organizations were to be consulted when decisions were made on distributing relief aid and on other plans on relief work.[187]

So who got relief grain? Using the figures cited above on evacuation, some 220,000 people were evacuated from Jinghai and Dakang and the housing statistics indicate that 203,829 people were actually housed in the suburbs and the city proper. However, only 169,600 people received relief aid.[188] The reason for such big discrepancies in numbers would need an explanation. The city authorities may have reported the number of refugees upward in order to receive more relief aid from the central government. Some of it may also be explained by the exclusion of people with a 'bad' class background from relief aid and simply by the poor quality of statistical work, as well as the food some refugees brought with them. Police stations were made to enforce the regulation that those who could support themselves were not given relief grain. For example, when inspecting the situation of the refugees there, Nankai District Wandezhuang Police Station discovered that seven families who had enough food for themselves were asking for the government's help. A refugee named Fan Qingyan from Jinghai County and his family had taken 400 *jin* grain with them when they had been evacuated, but they had told the authorities that they had nothing to eat and applied for food allowance. Refugee families could also lie about the size of their families to get more food.[189] These factors may account for the fact that about 50,000 refugees did not receive government food aid.

The city residents also played a role in feeding and clothing refugees. The city asked its residents to donate aid to the refugees by mobilizing a donation campaign that lasted to the end of December 1963. The City Government organized a propaganda campaign to study the heroic self-diminishing model of the people from water diversion areas and to support the disaster areas' resumption of production and reconstruction work. As instructions from the City Government in late October noted, city people should respond to the refugees' hardship in the spirit of 'people in the same boat helping each other' and to study the heroic self-effacing example of the people from diversion areas.[190] It was emphasized how helping such heroic refugees was the unwavering duty of urban residents.[191] Under this policy, in November and December city residents donated 9.6 million *yuan* of cash and clothes, shoes, cotton cloth, firewood, food grains and utensils to the refugees.[192]

However, voluntary donations were not enough. To improve refugees' nutrition, it was further ordered in November that people in city districts start to donate their monthly 0.5 *jin* (*c.* 250 g) of extra nourishment beans to the refugees until June 1964. As we have seen, these extra rations had been given to urban residents during the GLF famine in early 1961. It was further instructed that those who had more than enough food to eat could donate even more, while those who had less could, in principle, abstain from donating. *Danwei* leaders were instructed to carry this policy out fully and practical problems were to be reported to the Grain Bureau.[193] Altogether some 5 million kg of grain was collected this way.[194] For once, the urbanites were feeding the peasants.

This decision seems to be related directly to Chairman Mao who was well aware of the flood situation in Hebei and visited the province personally at least twice during late summer and autumn 1963. During his first visit on 11 October, Mao's special train stopped at Handan city (to the southwest of Tianjin) and he discussed with Provincial Secretary Lin Tie and Governor Liu Zihou the flood situation. Mao was briefed on the damage, relief work situation, refugees and public sentiments. He emphasized adequate food provisions for victims. Then on 12 November, Mao was returning from a tour in the provinces and stopped in Tianjin. He once again met Lin Tie and Liu Zihou, who gave a work report to Mao, including on the relief work situation in Hebei. In the discussion, Mao raised the issue that the daily 5 *liang* grain ration for refugees was not enough.[195] It is hardly a coincidence that the drive to donate urban rations to refugees was ordered after the visit. High-level intervention had helped Tianjin many times during the GLF famine and the same also applied in 1963 although this time lobbying was done by provincial leaders and benefits went to refugees.

Central relief grain was also made available in the city. Up to the end of 1963, the central government provided some 5 million kg of grain to the refugees as well as other foodstuffs and necessities in Tianjin,[196] as part of much larger assistance to Hebei Province. By the end of 1963, 10.38 million kg of relief aid in grain from different sources and 1.19 million *yuan* in cash had been used for relief aid, providing nutrition for 169,600 people and clothing for 47,000 in Tianjin.[197] Local army units conducted their own campaign to donate money and living materials to the refugees,[198] and suburban cadres mobilized communes to donate grain to refugees.[199]

The central government also provided special funds for relief expenses, clothing, cotton and the housing of refugees. The City Government issued orders on the proper use of disaster relief and reconstruction funds. A degree of distrust in grassroots cadres is visible in the instructions as the code of conduct included banning corruption, using funds for personal purposes, spending recklessly, or diverting money to other uses. Further, it was emphasized that the funds should not be used to benefit cadres' relatives or friends, and cadres were to abstain from making wilful and dictatorial decisions when using the funds. The Financial Bureau and the Civil Affairs Bureau had to supervise how the funds were used following the principle of 'four cleans', i.e. good management, distributing funds properly, recording all transactions on paper, and archiving them.[200]

Making the urbanites 'volunteer' to donate their grain rations to refugees was not the only donation drive in the city. There were a number of other initiatives, which showed how the City Government was using all its bureaucratic imagination to solve the problem. In December, the City Government ordered residents to start providing refugees with supplementary foodstuffs in order to alleviate the shortage of grains. The City People's Government decided that suburban peasants should supply refugees with dried vegetables from the autumn harvest. Parts of chopped or otherwise damaged vegetables could also be used after proper screening and drying, and distributed according to the instructions from the City Disaster Relief Committee. City districts were assigned their own quotas for providing these dried vegetables with instructions to strive to surpass them.[201] As we have seen, such dried vegetables had also been used during the GLF famine to supplement and augment grains. Their nutritional value was dubious, but probably they were better than nothing.

Refugees were also remembered during Chinese New Year in mid-February 1964. The provincial orders instructed that Hebei peasant refugees be given 250 g of extra rations and fine flour in order to make *jiaozi* during the Spring Festival.[202] However, 'urban refugees' as the ones living in Tianjin were called, were given 2 *jin* of grain, 1 *liang* of oil and 4 *liang* of pork extra for the festivities.[203]

Importantly for the relief effort, refugees were also given free medical care. This was the responsibility of the city Health Bureau, but medical costs were at least partly covered by special funds provided for this purpose from the central government.[204] By the end of 1963, free disaster-relief medical services were used 800,000 times by refugees.[205]

However, already during the early emergency mobilization in August, it was clear that relief for nothing was not the basic policy. Instead the basic Maoist social policy principle of making people work for their own relief was also followed in this case. The Party Centre laid down this principle by instructing that 'Relying on collective effort, using production self-help [*shengchan zijiu*] as the mainstay and the state's relief aid as supplement' was the basic policy to be used in relief work and reconstruction.[206] The policy was therefore to make refugees support themselves as much as possible, which had its roots in a similar policy in imperial times. The City Government instructed that those refugees who had brought raw materials with them should engage in sideline production (*fuye*). City bureaus had to buy these sideline products and distribute them in order to encourage other refugees to produce as well. Refugees also had to be provided with raw materials to encourage them to expand production activities.[207] By the end of 1963, some 28,000 refugees had been organized to produce over 30 different items as sideline production.[208]

Apart from producing sideline products, it was instructed that 'young and robust' refugees should work for a salary, and the Labour Bureau should arrange for temporary work permits for them to alleviate their problems and relieve the state's burdens. In this work the class line also had to be carried out, as it had to rely on poor and middle peasants and their mutual help.[209] It appears that this policy was not that efficient, as by the end of 1963 the Labour Bureau had assigned only 7,800 refugees to temporary jobs, such as constructing earthworks and collecting autumn crops.[210]

Encouraging sideline production was nevertheless politically sensitive. If such operation was assigned to too low levels of operation, i.e. to households and individuals, it could come too close to capitalism. As seen in Chapter 1, Chairman Mao had just a year before prevented the application of household responsibility systems in agricultural production where the principle had been to try to make households responsible for production decisions and Tianjin's authorities had also zigzagged in their policy on peasant markets. Promoting sideline production invoked this ideological problem once again. Probably this is why the City Government was careful to instruct that while self-help through sideline production was also the mainstay in reconstruction, collective production had to be its mainstay, and household production only secondary. The principle was that large-scale production belonged to the collectives, while only scattered small-scale production that could be done by households could be distributed among the commune members.[211]

The problem was common to the Chinese command economy in general. While it sought to promote individual enthusiasm for production, too much, or wrongly induced, individuality was not welcome for ideological reasons. This was also seen in the way the city issued an order to curb companies from hiring refugees of their own accord. It was emphasized that it was the duty of refugees' counties of origin and city districts, as well as the City Labour Bureau, to arrange temporary work in a planned and comprehensive manner according to need. Companies were therefore ordered to stop hiring refugees by themselves.[212] This continued the line that had already begun during the GLF. It was understandable that companies would hire refugees when their best workers had been sent to maintain dykes and even after this when the city demanded increases in production and refugees provided a cheap and easily available source of extra manpower.

Refugees were also banned from engaging in peddling or market speculation,[213] which continued the struggle against black markets. The structures of command economy, namely rigid rationing of the labour force and controlling its mobility that had come into force by the beginning of the Great Leap Forward were therefore also enforced under the crisis situation, but the loophole of sideline production and collective economy was used to create incentives for refugees to work for their own relief.

There are some contradictory statements on how production self-help worked with the refugees. The City Government admitted in December 1963 that production self-help had not developed quickly enough, production methods were not versatile enough, and manpower was not fully engaged in production. However, the same document stated that because of promoting production, self-help refugees' living conditions were 'basically in order', their sentiments were stable, and they would use the winter to establish the basis for reconstruction of their homes.[214] The same principles were indeed carried out in the reconstruction of flooded areas in 1964.

The Reconstruction of Disaster Areas

The basic policy was to send refugees back home as soon as possible; their stay in Tianjin was to be a temporary one. This was shown in how propaganda was instructed to keep refugees' spirits up on the prospects of victory in the struggle and an early return to villages. Therefore, the refugees were told not to sell their cattle, their property or production tools while in the city.[215] This had clear production-based calculations behind it. Already in early August, the province instructed that in the areas where water had receded, the masses were to be organized so that they could resume production as soon as possible.[216] In late August, instructions ordered the repatriation as soon as possible of those from areas where the flood had already receded in order to make quick reconstruction of their homes possible.[217] According to the City Government reports, the larger repatriation of refugees began by mid-November by sending 23,000 refugees from Wujing and Anci to their homes.[218] In December, it was ordered that healthy refugees from areas where there were production possibilities should be repatriated, but the sick or those having otherwise real problems in going back should be allowed to stay for a while longer.[219] By the end of December 1963, 94,329 refugees had already returned home, while 79,378 were still left in the suburbs (and the rest were probably located in the city proper).[220] By December, at least 33,000 Dakang people had already been sent home. After arriving there, they had to engage themselves in rebuilding their own houses and repairing damaged ones. Official reports claimed that this repatriation progressed smoothly.[221] In Jinghai over 40,000 houses were rebuilt and over 70,000 repaired in winter 1963–4 .[222] The timing was ideal in the terms of production, as this was agricultural low season.

The Tianjin City Government announced production relief work (*chansheng jiuzai gongzuo*) in the form of rebuilding agricultural production in disaster areas as its priority for 1964. All city bureaus and *danweis* were instructed to participate in rebuilding. The City Construction Bureau was to participate in rebuilding agricultural production through providing machinery, tools and materials. The Commerce Bureau was to work on providing disaster areas with daily necessities, especially those that would help people to survive winter. These areas' sideline production also had to be promoted. Districts had to mobilize those areas with bumper harvests to sell more of their grain and vegetables to the state to support disaster areas. The Health Bureau had to continue to organize medical work teams and medicine to be sent to villages to prevent diseases there. Stronger ideological work to help people support disaster areas throughout the city and support the mutual help of cities and villages was also called for.[223]

In December, Hebei Provincial Health Department stressed the need to offer free healthcare services to the returning refugees.[224] Accordingly, the Tianjin Health Bureau organized three teams totalling 1,362 people to provide medical services and disease prevention in Jinghai and some Hebei Province areas. These consisted of small medical groups of 4–5 people and they were deployed to create a network of medical and disease prevention services in these areas. Their services were used 164,094 times in the spring of 1964. The teams also took part in pest eradication work in disaster areas, educating health workers, providing water sanitation, etc. The teams' assignment ended in June 1964.[225]

In rebuilding disaster areas, the bulk of the work was to be done by the peasants themselves under the principle of production self-help. However, the City Government, Finance Bureau, and other related bureaus organized help under the 'four priorities' (organizing production, supply, purchasing, and helping marketing).[226] In December, the City Government issued further instructions on production self-help under instructions from the Party Centre and the province. Now it was ordered that a sideline work tide be created for production self-help. It was declared that 'Developing side-line work and increasing investments [in it] is the way to mend the damage caused by the disaster, an important road to victory against the disaster and a compelling wish of the great masses'.[227] Specific conditions in the district had to be the basis of the policies. For example, in Dakang District, fisheries were to be developed along these lines. Also repairing river embankments was one occupation that was recommended for returned refugees. Marketing cooperatives were ordered to provide raw materials and purchase sideline products promptly.[228]

All city bureaus were ordered to engage in helping disaster areas to rebuild and help them in a united manner. Raw materials and work power were to be provided by the Labour Bureau and marketing cooperatives. Suburbs, communes and production brigades were instructed to let to the best of their abilities their refugees engage in sideline production. Production technology was also to be provided to them as far as was possible.[229]

The provincial and city authorities also decided that all administrative units, collectives and companies had to collect old and discarded materials and reduced-price or worthless goods to support disaster districts. These were collected from the whole city and included altogether 310,000 items, such as winter clothing and other usable goods, repair materials for houses and kitchen utensils. Over 70,000 *jin* of scraps of industrial material were also donated for reconstruction.[230]

Under this policy, production self-help was actively organized in disaster areas. In Jinghai County alone, some 90,000 people were engaged in sideline production for relief in 1964. A total of 177 different kinds of items were produced, and it was estimated that, by the end of 1964, disaster areas had benefitted by some 4 million *yuan* from this production. The Commerce Bureau also developed a system of selling these products on consignment, whereby agricultural sideline products were sold worth over 8 million *yuan*. 16,589 men were also hired for earthwork projects by various city bureaus.[231]

The city also provided funding for rebuilding. In the first half of 1964, Jinghai and Beidakang were given 9.2 million *yuan* for agricultural investment, repairing housing and living expenses. An extra 300,000 *yuan* was also made available for free medical services. Credit cooperatives gave loans for living expenses worth 1.7 million *yuan*, and for production investments 2.5 million *yuan* was given. Foodstuffs, firewood and other necessities were provided to Jinghai as relief aid. Officially, the living conditions were reported to be ‘basically’ in order in disaster areas by June 1964.[232] This was also when central reconstruction aid to Hebei ended.[233]

However, even in the midst of rebuilding peasants were not allowed to relax in their political education. The City Production Self-Help Committee instructed that refugees and disaster areas were not exempt from socialist education campaign. Instead the refugees were to organize poor, lower and middle peasant associations, regardless of whether they had already returned or were still located in their evacuation places, and to carry out the campaign and generally assist in reconstruction work.[234] Propaganda work, an integral part of Maoist disaster management, did not let up.

The Aftermath

Tianjin city was saved from the flood, but it had been a close call, and large parts of Hebei Province suffered a lot. Like Sage Yu several millennia before him, Chairman Mao become convinced that Man had to conquer nature and tame the waters. In his visit to Hebei in October, he inspected a map of the planned preventive waterworks in the province. During his brief second visit in November, he noted that the waters must be tamed even if it took over ten years. He might not be there to see it finished, but it should still be done by this generation, he instructed his audience.[235] In the same month Mao gave 'great instructions' to 'Bring the Hai River under permanent control' in order to change the hundreds of years of history of floods in the region.[236] Liu Shaoqi gave instructions to the same end noting: 'People in the province unite! Exert your powers conscientiously and determinedly to solve the problem within a timeframe of some 20 years, build up the irrigation works of the Haihe stage by stage.'[237]

Authorized in 1965 by the State Council, the plan involved 300,000 men per year engaged in the new Haihe Project.[238] It involved constructing 85 large and 1,500 small water reservoirs along the Hai River tributaries. These could contain more water than had been involved in the whole 1963 flood. Moreover, ten hydropower stations were constructed in the hilly districts upstream. Altogether 3,400 km of rivers were dredged and 4,300 km of flood embankments were erected. This project increased the discharge capacity of the five rivers in the Haihe drainage area tenfold. In essence, the Hai River's main tributaries became canals or were redirected and given their own outlets to the sea. It took 14 years to complete the project by 1979, but it basically ended the history of the Haihe flooding its surroundings for good, dramatically reducing flooding in the region in general.[239]

Conclusions

The disaster management cycle for the 1963 Great Flood ended in the summer of 1964. Compared to the GFL famine crisis management, Tianjin rallied in 1963–4. The city was saved by trading a major disaster for a manageable refugee and reconstruction problem. Refugees were taken care of in a fairly basic, but still adequate, way by the standards of the time. In this campaign, the Maoist state demonstrated probably its most efficient aspects and strengths in protecting its citizens from a disaster, when it had just demonstrated its greatest weaknesses in the same task.

Disaster management of the Tianjin flood in 1963 shows how the Maoist sociopolitical structures could be effectively used to protect the lives and property of urban citizens. Urban and rural work units acted as generic mobilization units, which meant that they could be quickly transformed from specialized production units into flood combat outfits with relatively little friction in the process. This effect was further strengthened by the GLF innovation of militias. The same structures could also be relied on in the later stages of the disaster management cycle to contribute to disaster relief. In general, however, success or failure in disaster management under the Maoist regime also depended on several other factors that were related to the way the political system operated.

The key difference between 1959–61 and 1963 was that in 1963 neither central nor local officials had any reason to resist emergency measures and relief effort. The central government demanded that the city be saved from the flood and monitored disaster management closely. The flooding of the city would have benefitted no-one at local level and, in any case, could not have been concealed from central government, it being too closely located to Beijing. Local officials were likely to face punishments should they fail to prevent such a large-scale disaster from occurring, and commendation should they succeed in preventing it, as they did. Moreover, the reason for the emergency was clear to all: extraordinarily excessive rain that had seemingly no man-made aspects, especially because the collapse of reservoirs that caused the two flood peaks was kept a secret from the public. Indeed, this was an opportunity for the party-state to prove its capability in disaster management after failure in the GLF. Under these circumstances, the pre-existing planning and preparations, and the skill of mass mobilization could be utilized to the full.

There is also the question of how the GLF influenced Tianjin's ability to respond to the flood. Propaganda claimed that the victory was based on the correct policies since the 1950s and the GLF.[240] It was claimed that reservoirs built during the GLF stopped two-thirds of the floodwaters and that the great Haihe sluice built in 1958 prevented sea water from coming up against floodwaters in the river, thus increasing drainage capacity greatly.[241] However, the GLF achievements were actually mentioned only afterwards. During the emergency mobilization, the Party did not re-raise any of the slogans of the failed GLF that was very fresh in people's memories. The accuracy of these claims is therefore difficult to ascertain in the case of Tianjin, but contemporary research has noted how many ill-conceived waterworks constructed during the GLF were actually conducive to flooding.[242] Some of the most valuable elements of waterworks that helped to saved Tianjin, such as Duliujian, had actually been constructed before the GLF and left to decay during it. However, the GLF did leave Tianjin with added organizational readiness in the form of the people's communes and militias and both of them undeniably played a positive role in flood prevention work.

Another aspect of Maoist politics and priorities was also evident in the orders given to the home front, and what was missing in them. The way the campaign was conducted suggests that the motivation behind central government's concern over Tianjin was clearly not humanitarian; rather it was about how to keep production going under crisis conditions. Related to this, there is no indication of any plans for the evacuation of the city if the dykes had not held. If there was one, it was not made public, which rendered it almost meaningless. While propaganda painted a picture of a city where factories were running, offices working and shops kept open as normal even during the campaign against the flood,[243] residents were not instructed what to do and where to go if water suddenly did rise in the city streets. A large part of disaster-related reporting in the *Tianjin ribao* was useless propaganda that did nothing to prepare people to survive if the worst happened. It is as if the possibility of the city drowning never occurred to the city and provincial leadership as a real option that had to be included in planning. As the thought must have crossed their minds, they were either very certain of victory or decided to gamble all-or-nothing. Such a decision tells much about the priorities of the Maoist state, where risk to human lives was ultimately always a lesser concern than production figures. There indeed was a 'cult of production' that set the basic parameters for Maoist disaster management.

Notes

1 Li (2007, Chapter 1) describes the geography of Hebei plains in detail.
2 http://www.britannica.com/EBchecked/topic/251717/Hai-River-system?anchor=ref158256 (accessed 6.5.2013).
3 FXZL 25.9.1963a, 101.
4 FXZL 21.6.1963, 3.
5 FXZL 30.5.1963, 7.
6 FXZL 25.5.1963, 1–2; FXZL 21.6.1963, 3–6.
7 FXZL 21.6.1963, 4.
8 Ibid., 5–6.
9 FXZL 30.5.1963, 11.
10 TJZHB 11 (29.6.1963), 5.
11 Ibid., 3.
12 Ibid., 4.
13 Ibid., 3–4.
14 Ibid., 4–5.
15 Liu Shaojin 2004, 80–82
16 FXZL 24.7.1963, 26.
17 Ibid., 27.
18 FXZL 25.9.1963a, 101.
19 MZHJZH 1995, 246
20 Liu 2004, 83.
21 Li 2007, 367.
22 Tianjinshi fangxun zhihuibu (1963/1999b), 45.
23 MZHJZH 1995, 248.
24 Liu 2004, 84.
25 TR 9.8.1963, 2.
26 TR 27.8.1963, 2.
27 TR 9.8.1963, 2.
28 This situation was brought about by 30 small reservoirs and one medium-size reservoir breaking their embankments in the Taixingshan mountain districts of Hebei. FXZL 8.8.1963, 29.
29 Ibid.
30 Liu *et al.* 1964, 30; Tianjinshi fangxun zhihuibu (1963/1999a), 43.
31 FXZL 8.8.1963, 30–32.
32 Ibid.
33 Tianjinshi fangxun zhihuibu (1963/1999a), 43–4.
34 His name was Teng Haiqing (Liu 1964, 37).
35 Wan Xiaotang's active leadership in the campaign against the flood is noted by Li Ding 2001, 49.
36 MZHJZH 1995, 248.
37 FXZL 8.8.1963, 33–34.
38 FXZL 15.10.1963a, 190.
39 These were the political outcasts of Maoist society: the landlords, rich peasants, counterrevolutionary elements, bad elements and rightists.
40 FXZL 15.10.1963a, 190–1.
41 FXZL 13.8.1963a, 35–6; later sub-FPHQs would report that there was no 'political criminal activity' detected during the struggle. FXZL 25.9.1963a, 104.
42 FXZL 30.9.1963a, 116.
43 TR 9.8.1963, 2; TR 11.8.1963, 1.
44 Liu *et al.* 1964, 8 and 31–2.
45 Wei *et al.* 1964, 2.
46 MZHJZH 1995, 248.

47 FXZL 15.10.1963a, 191.
48 FXZL 8.12.1963, 42–3.
49 Liu *et al.* 1964, 33.
50 FXZL 13.8.1963, 36; TR 14.8.1963, 1.
51 Liu *et al.* 1964, 34; FXZL 30.9.1963, 111.
52 FXZL 30.9.1963c, 89–91.
53 Ibid. 91–92.
54 FXZL 25.9.1963a, 101–2.
55 FXZL 30.9.1963c, 92.
56 Ibid.
57 Ibid.
58 Ibid.
59 Ibid., 94.
60 Ibid., 92–4.
61 FXZL 25.9.1963a, 102.
62 FXZL 15.10.1963a, 191.
63 TR 29.8.1963, 2.
64 FXZL 15.10.1963b, 121–4.
65 FXZL 25.9.1963b, 133.
66 TR 7.9.1963, 1.
67 Liu *et al.* 1964, 39.
68 FXZL 25.9.1963b, 133–6.
69 FXZL 30.9.1963c, 95.
70 TR 3.9.1963, 1.
71 TR 22.9.1963, 1; MZHJZH 1995, 248–9; FXZL 30.9.1963b, 138–9.
72 Liu *et al.* 1964, 44.
73 TR 7.9.1963, 1.
74 FXZL 15.9.1963.
75 FXZL 15.10.1963a, 191.
76 FXZL 15.9.1963.
77 FXZL 1.9.1963, 209.
78 TR 27.9.1963, 1.
79 TJJS 1995, 510.
80 FXZL 15.10.1963a, 191.
81 Tianjinshi fangxun zhihuibu (1963/1999a), 44.
82 Tianjinshi fangxun zhihuibu (1963/1999b), 46.
83 TR 18.8.1963, 1.
84 FXZL 15.10.1963a, 192.
85 Ibid., 192.
86 FXZL 30.9.1963a, 115–16.
87 Ibid.
88 There was also a reason to remain vigilant. According to statistics, between 20 August and 8 September, 98 serious and 126 dangerous situations occurred on embankments. All these were solved through rapid action. Tianjinshi fangxun zhihuibu (1963/1999b), 47
89 TR 16.8.1963, 1
90 TR 21.8.1963, 1; TR 25.8.1963, 1
91 Ibid.
92 Tianjinshi fangxun zhihuibu (1963/1999b), 47.
93 TR 21.8.1963, 1; TR 16.8.1963, 1.
94 E.g. TR 29.8.1963, 2.
95 TR 27.9.1963, 1.
96 Tianjinshi fangxun zhihuibu (1963/1999a), 43–4.
97 FXZL 8.8.1963, 31.
98 FXZL 16.8.1963, 208.

99 Wang *et al.* 1964, 64.
100 FXZL 30.9.1963a, 117.
101 FXZL 15.10.1963b, 122–3.
102 TR 21.8.1963, 1.
103 TR 16.8.1963, 1.
104 TR 18.8.1963, 1; Tianjinshi fangxun zhihuibu (1963/1999b), 46–7; Tianjinshi fangxun zhihuibu (1963/1999c), 50.
105 Liu *et al.* 1964, 35.
106 TR 15.8.1963, 3.
107 TR 24.8.1963, 2.
108 TR 21.8.1963, 1.
109 TR 15.8.1963, 3.
110 Tianjinshi fangxun zhihuibu (1963/1999b), 47.
111 Tianjinshi fangxun zhihuibu (1963/1999c), 50.
112 FXZL 16.8.1963, 208.
113 FXZL 28.9.1963, 264–5.
114 FXZL 23.8.1963, 212.
115 TJZHB 11 (29.6.1963), 5.
116 TJZHB 16 (31.8.1963a), 3.
117 Wei *et al.* 1964, 1.
118 Ibid., 10.
119 E.g. TR 15.8.1963, 3; TR 2.9.1963, 1.
120 TR 19.8.1963, 1.
121 TR 24.8.1963, 1.
122 TR 15.8.1963, 3.
123 TR 16.8.1963, 1.
124 Li *et al.* 1964, 50.
125 TR 18.8.1963, 2.
126 Wang *et al.* 1964, 64
127 TR 17.8.1963, 2.
128 TR 14.8.1963, 2.
129 Liu *et al.* 1964, 46.
130 TR 13.8.1963, 1; TR 13.8.1963, 1.
131 Wei *et al.* 1964, 18.
132 For workers e.g. TR 19.8.1963, 1.
133 E.g. in TR 29.8.1963, 2 and TR 29.8.1963, 2.
134 TR 24.8.1963, 1.
135 TR 15.8.1963, 3.
136 TR 21.8.1963, 2.
137 TR 18.8.1963, 2.
138 Tianjinshi fangxun zhihuibu (1963/1999b), 46.
139 Tianjinshi fangxun zhihuibu (1963/1999c), 50; TJZHB 16 (31.8.1963a), 2.
140 TR 21.8.1963, 2.
141 TR 21.8.1963, 1.
142 TR 21.8.1963, 2.
143 Tianjin Ribao reporters 1964, 112.
144 Tianjin Ribao reporters 1964, 112–13.
145 Tianjinshi fangxun zhihuibu (1963/1999b), 46–7; Tianjinshi fangxun zhihuibu (1963/1999c), 50; TR 4.9.1963, 1.
146 TJZHB 16 (31.8.1963a), 3.
147 FXZL 23.8.1963, 212.
148 TJZHB 16 (31.8.1963a), 3.
149 Liu *et al.* 1964, 42–3.
150 TR 25.8.1963, 1.

151 TR 25.8.1963, 1; TR 17.8.1963, 2.
152 TR 18.8.1963, 2.
153 Wei *et al.* 1964, 3.
154 TJZHB 23–24 (31.12.1963), 8; TR 21.8.1963, 2; TR 27.8.1963, 2,
155 TR 6.9.1963, 5; TR 18.8.1963, 2; TR 21.8.1963, 2.
156 TR 25.9.1963, 1.
157 TR 18.8.1963, 2.
158 TR 25.9.1963, 1; TR 21.8.1963, 2; Li *et al.* 1964, 134–46.
159 TR 25.8.1963, 2.
160 TR 25.9.1963, 1.
161 HBMZHJ 2010, 371–2.
162 FXZL 1963, 369–70.
163 FXZL 22.9.1963, 227.
164 Ibid., 226
165 Ibid., 226–7.
166 FXZL 3.10.1963, 349.
167 MZHJZH 1995, 247–51; TJJS 1995, 509.
168 FXZL 10.8.1960.
169 *Jinghai xianzhi* 1995, 510.
170 TR 8.9.1963, 1.
171 TJZHB 16 (31.8.1963b), 4,
172 MZHJZH 1995, 250.
173 Ibid.
174 TJZHB 16 (31.8.1963b), 4. Originally, a distinction was made between peasants from flood diversion areas and other refugees, but in the end a great majority of refugees, 98.5per cent (MZHJZH 1995, 255), were of the first category and thus the distinction probably did not matter much.
175 *Jiuzai jianbao* 15 (8.9.1963), 39.
176 Ibid.
177 TJZHB 16 (31.8.1963b), 4–5.
178 HBMZHJ 2010, 372.
179 Ibid.
180 MZHJZH 1995, 250–1.
181 TJZHB 16 (31.8.1963b), 4–5.
182 MZHJZH 1995, 251; TJZHB 16 (31.8.1963c), 5.
183 TJZHB 21–22 (16.12.1963), 3.
184 TJZHB 21–22 (16.12.1963), 4–5.
185 TJZHB 16 (31.8.1963c), 5.
186 Dixon (1981) explains these principles in detail.
187 TJZHB 21–22 (16.12.1963), 7.
188 *Caishuizhi* 1995, 371.
189 *Jiuzai jianbao* 9 (31.8.1963), 23.
190 TJZHB 19 (1.11.1963a), 10.
191 TJZHB 21–22 (16.12.1963), 6; MZHJZH 1995, 254.
192 MZHJZH 1995, 254.
193 TJZHB 19 (1.11.1963a), 10.
194 The campaign was ordered to last until June 1964, but according to the gazetteer, it lasted only to March 1964 (MZHJZH 1995, 254).
195 Yuan 2010, 1269–70 and 1278–9.
196 MZHJZH 1995, 252.
197 MZHJZH 1995, 252; *Tianjin jianzhi*, (TJJZ 1991, 1199) notes that ‘the state’ not the ‘city’ provided this relief aid.
198 TJJS 1995, 510.
199 MZHJZH 1995, 252.

200 TJZHB 21–22 (16.12.1963), 5.
201 TJZHB 21–22 (16.12.1963), 5.
202 TCA 29.1.1964.
203 TCA 11.1.1964.
204 TJZHB 16 (31.8.1963c), 5.
205 MZHJZH 1995, 253.
206 TJZHB 21–22 (16.12.1963), 7.
207 TJZHB 16 (31.8.1963c), 5.
208 *Caishuizhi* 1995, 371.
209 TJZHB 16 (31.8.1963c), 5.
210 TJZHB 21–22 (16.12.1963), 3.
211 Ibid., 4.
212 TJZHB 19 (1.11.1963b), 7.
213 TJZHB 19 (1.11.1963), 4.
214 TJZHB 21–22 (16.12.1963), 3–4.
215 TJZHB 16 (31.8.1963), 5–6. However refugees were allowed to sell smaller farm animals with them.
216 FXZL 10.8.1963.
217 TJZHB 16 (31.8.1963c), 5
218 TJZHB 21–22 (16.12.1963), 3.
219 TJZHB 21–22 (16.12.1963), 6.
220 MZHJZH 1995, 251.
221 TJZHB 21–22 (16.12.1963), 3.
222 *Jinghai xianzhi* 1995, 510.
223 TJZHB 23–24 (31.12.1963), 10.
224 TCA 401206800-X0283-C-001370-015 (28.12.1963).
225 MZHJZH 1995, 253.
226 MZHJZH 1995, 253.
227 TJZHB 21–22 (16.12.1963), 4.
228 Ibid.
229 Ibid.
230 MZHJZH 1995, 254.
231 Ibid., 253.
232 Ibid., 253.
233 HBMZHJ 2010, 372.
234 TCA, 1.15.1964.
235 Yuan 2010, 1269–70 and 1278–9.
236 TJZHB 23–24 (31.12.1963), 9.
237 Wei *et al.* 1964, 10.
238 Li 2007, 367.
239 MZHJZH 1995, 255–6; Li 2007, 367; http://www.britannica.com/EBchecked/topic/251717/Hai-River-system?anchor=ref158256
240 Li 2007, 4.
241 Wei *et al.* 1964, 4–5.
242 Dikötter 2010, 178–83.
243 Liu *et al.* 1964, 42.

3 The Great Earthquake of 1976

Apart from floods, Tianjin has a long history of earthquakes. Between 1345 and 1990 there were all together 1,902 recorded tremors in the city, 359 of which were small (M 3–5) and 18 mid-strength (M 5–7).[1] Among disasters, earthquakes are known as 'mass harmers', since they usually strike without warning and can cause serious damage to large populations and infrastructure in a very short time. This was also true on 28 July 1976 when Tangshan Prefecture adjacent to Tianjin was hit by an M 7.8 earthquake, which also severely affected Tianjin. Indeed, the city suffered the greatest damage in its recent history of natural disasters, although the earthquake does not carry 'Tianjin' in its name. However, before analysing disaster management related to the earthquake, a disaster of another kind, which formed the political background to the Tangshan earthquake disaster management in the city, needs to be shortly addressed.

In Chinese contemporary history, the period 1966–76 signifies the Great Proletarian Cultural Revolution (CR). It was launched by Chairman Mao in order to purge his real and imagined enemies from the Party and cleanse the party-state of 'revisionists' and 'capitalist roaders'. The campaign itself is beyond the scope of this book, but it did directly affect disaster management. The Ministry of Interior was abolished in 1969 and its disaster management functions were dispersed to other ministries, which at the time were in disarray because of the CR.[2] Although central government had mostly recovered by 1976, it took until 1978 for the Ministry of Civil Affairs to be (re)established.

The CR affected also Tianjin's city administration. In Tianjin, as all over China, the CR began as a campaign in cultural circles in spring 1966, but soon evolved into a campaign against 'capitalist roaders' throughout the Party. University and high school students were activated on city campuses in June and began to attack their teachers and leaders. In late June, Tianjin high school students also began openly to criticize city leaders, targeting the head of the Education Department Zhang Huaisan, but also Wan Xiaotang. At the end of August, Red Guards stormed the city's Party Headquarters and surrounded the building. Wan Xiaotang faced the demonstrators and argued with them that the CR had to be held under party control. This did not go down well with the Red Guards. On 16 September, the Red Guards from No. 16 Middle School – the hotbed of Red Guard activism in Tianjin – and other Red Guard groups convened a mass meeting to criticize city leaders, including Wan Xiaotang. A second mass meeting was held on 19 September. During it, Wan Xiaotang succumbed to a heart attack and died. Wan was mourned openly by city residents.[3]

The Party Centre dispatched Xie Xuegong from the Huabei Office to Tianjin on November 1966 to take control of the Tianjin Party. In January, following the example of Shanghai, Tianjin was enveloped in power seizures and struggles between rebel and conservative Red Guard groups and mass organizations. The City Party Committee was toppled on 18 January, and the party organization virtually ceased functioning. However, Xie Xuegong used the situation to his benefit and established a cadre committee to seize power in the city for the creation of so-called revolutionary committees, and directed a purge against the 'Wan (Xiaotang) and Zhang (Huaisan) counterrevolutionary clique', targeting many old cadres in the party and city administration including Song Jingyi and Li Gengtao. The party purge, which the CR was largely about, was therefore a top-down job in Tianjin. The Tianjin Revolutionary Committee, which replaced the City Government, was established in December 1967 under Xie, but it took until 1971 to revive the entire party structures in the city. After this, the city administration did not witness major organizational changes again until 1978.[4] It was this 'revolutionized' administration that faced the earthquake in July 1976.

One additional important administrative change took place in January 1967, when Tianjin regained its status as a municipality directly under central government. The city and Hebei Province parted company administratively, which was probably good for the city's development in the long run. However, for the first ten years or so after the decision, first the CR and then the Tangshan earthquake prevented the city from utilizing its regained higher administrative status to the full.

Earthquake Preparedness

Tianjin City authorities were aware of the danger of a major earthquake hitting the region one day. Therefore, like floods, the city had been building up preparedness for them. Coincidently, city records trace the beginning of earthquake preparations to 1966. That year, a strong earthquake shook Hebei Xingtai County leaving about 8,000 persons dead. This was the first major earthquake to strike China after the founding of the PRC. After visiting the disaster site, Premier Zhou Enlai issued instructions on improving the earthquake-forecasting system and preventive measures. In the anarchy of the CR, it took five years to establish National Earthquake Administration under the State Council for predicting and monitoring earthquakes as well as for improving earthquake preparedness and emergency responses.[5] Following the central government's instructions, Tianjin also set up a Headquarters for Resisting Earthquake and Relief Work (RERWHQ), but this no longer existed in 1976.

Before 1976, the organizational structure of the city earthquake preparedness work went through much restructuring and development. An important boost to work was the M 6.3 earthquake that rocked the city in March 1967 leaving 284 dead, nearly 20,000 people homeless and halting production in the city for a while. Now every government office established a special earthquake preparedness organization. Planning emergency activities in the case of an earthquake was developed, forecasting activities strengthened, special rescue teams set up and buildings' structural ability to withstand earthquakes was surveyed. After this, earthquake work organizations were restructured several times before 1975 when they reached more or less the form they had in 1976.[6]

Reorganization in 1975 was due to the M 7.3 earthquake that struck Liaoning in February 1975. This made central government repeat its warning to Tianjin on the probability of a major earthquake in the region in the near future. The same month, the City Party Committee decided to establish the Resist and Repel Earthquake Leading Small Group (*Fangzhen kangzhen lingdao xiaozu*), which then convened a city-wide leading cadre meeting on earthquake and emergency rescue work. It was decided that the entire city would be mobilized for earthquake work in order to mitigate disaster in the case of a major earthquake. This meeting also decided to establish the Tianjin City Resist and Repel Earthquake Office under the City Construction Committee to manage anti-earthquake construction projects.[7]

The meeting also decided on evacuation areas for people in case of an earthquake. These were mainly major streets and other open areas. It was also decided to prepare food, medicine, water, reed mats, bamboo and canvas for tents for emergency lodging. First-aid teams were to be made ready, and emergency rescue teams were to be organized under a three-level (city, district, street) command structure. At city and district level, special 'remove hazards and rush-repair teams' were to be set up for repairing roads and bridges. The Transport Bureau was to set up special transport teams for materials and the injured, the Post and Telecommunications Office a rapid action team for repairing telecommunications, and the Construction and Building Management system's organizations set up teams for clearing dangerous buildings. It was further decided that the military war preparedness command and communication systems could be utilized under emergency conditions to lead rescue and relief operation. Small city, district and county earthquake leading groups were to have military members in them.[8] Under the conditions of the recent CR, this should have appeared quite normal anyhow.

In May 1976, the City Party Committee further decided to organize the City Earthquake Office (*Dizhenju*, CEO) to coordinate all anti-earthquake measures in the city. The meeting also declared general policy principles for earthquake work: limiting the number of victims, conducting propaganda work in order to raise residents' awareness of earthquakes, continuing the survey work conducted in unison by the experts and members of the masses, and improving the management of newly-built buildings for their ability to withstand earthquakes.[9] Earthquake offices were also established at office, district and county levels. This meant that both the Party and the City Government had their respective offices/small groups designated to guide anti-earthquake work.[10]

Further development of earthquake preparations was inspired by the 1975 central government instructions entitled 'Jing-Jin Region Criteria for Inspecting

Industrial and Public Buildings', referring to the capital and Tianjin. Under these criteria, city bureaus and *danweis* carried out surveys of their public, industrial and residential buildings with a view to their capability to withstand scale VII damage from an earthquake.[11] Based on the survey work, in late May 1976 the CEO published the 'Preliminary Plan on Earthquake Disaster Conditions and Resisting and Protecting against Earthquakes'. As the name indicated, this was a comprehensive plan on organizations needed and measures to be taken in a major earthquake situation. It laid out the responsibilities of all bureaus, offices and the army in the eventuality of an earthquake.

The plan outlined how, in case of an earthquake, there should be a Temporary Earthquake HQ (*Lin zhen zhihui bu*) that would consist of members of the General Office of the City Party Committee, the Military Preparedness Office, the Propaganda Office and the CEO, and other bureaus. It would take charge of the leadership of emergency rescue, evacuation, relief, and reconstruction work after an earthquake. Furthermore, all administrative levels down to streets and villages had to establish their own Earthquake Preparedness Offices to coordinate earthquake protection measures. City districts were instructed to draft their own similar plans under the leadership of small groups. All *danweis* were to organize their specialized teams and train them in their tasks. The same plan also outlined civilian and military sectors' responsibilities in the event of an emergency. These followed their respective functions and did not vary much from the duties outlined in 1975.[12]

As a general principle, it was ordered that when an earthquake warning had been given, all residents were to move outside. Factories had to gradually close down. Ships had to cast off and move out to sea. All members of task forces had to report to their posts, wait for orders, and prepare to move out quickly to rescue sites after an earthquake. Members of earthquake-forecasting stations and the people's earthquake stations also had to stay in their posts and continue reporting.[13]

As for immediate repairs and reconstruction after an earthquake, it was ordered that railroad traffic, electricity and water supply had to be resumed as soon as possible. Harbours had to be cleared of debris in order to resume the domestic and foreign flow of goods.[14] Three types of goods were to be stored for emergency purposes: those used in earthquake protection work, medical supplies and living necessities, but overstocking and waste was not allowed. Hazardous substances also needed to be surveyed and their management planned. Flammable goods had to be stored carefully. Railroad cars and engines also had to be deployed in a dispersed manner to increase their safety.[15]

Planning shows a good level of awareness by the city administration of measures necessary to mitigate damage in an earthquake. However, it is unclear how systematically different districts and units followed the plan, and how well they were prepared when the Tangshan earthquake hit. The districts' own earthquake protection plans based on the city instructions indicate that grassroots level units had months or even years to implement the policy depending on location.[16] Therefore, organizing and training different groups outlined in the plans was at best underway on 28 August. When the earthquake struck, all units down to street and production team levels did organize their own headquarters to resist earthquake and relief work, but whether they had existed already before the earthquake is usually not mentioned in the sources.

Apart from organizational measures, physical preparations formed a part of Tianjin earthquake preparedness. Here the 'Jing-Jin Region Criteria for Inspecting Industrial and Public Buildings' prompted a large survey of the condition of public, industrial and residential buildings in Tianjin. The results were rather discouraging as outlined in the 'Preliminary Plan on Earthquake Disaster Conditions and Resisting and Protecting against Earthquake' by the CEO. It was estimated that in a big earthquake (of category VII) some 27.1 per cent of public and 14 per cent of industrial buildings would suffer damage, and city infrastructure would be widely damaged. Furthermore, many city roads were narrow and old buildings could collapse on them causing traffic to be blocked. The danger of secondary damage was also high. Tianjin had a 'very unideal' location for many industries with flammable and toxic chemicals near residential areas. Therefore, fires causing severe secondary damage were likely, it was envisaged.[17]

There had been some measures taken to reinforce physical structures in the city. This had been prompted by the 1967 earthquake, but, more recently, in 1975 the State Construction Committee document no. 406 had ordered reinforcement of units that were important for production and 'political economy' in the city. There were all together 29 such units in the city, such as the City Party Committee and Government buildings, which were reinforced before the earthquake and were able to withstand it. However, only a few hundred residential houses were reinforced before the earthquake.[18] It is also clear that there had been not enough time or funds to carry out reinforcement even in many of the priority sites when the earthquake struck.

A separate, but integrated part of preparedness was the short-term prediction and reporting work, which is probably the most indigenous feature of the Maoist approach to disaster management.[19] First of all there were scientific geological survey stations that were assigned the duty of earthquake forecasting and warnings. The first of these had been established in July 1969 in Tianjin. In 1975 three more stations were set up in Jinghai, Ninghe and Ji Counties. Furthermore, the Oil Industry Bureau helped to drill a 1 km-deep hot water well at Dakang oil field to help in earthquake water chemistry surveys.[20] In addition to such natural scientific seismographic work, the people were also engaged in earthquake survey and forecasting work through the so-called people's survey and reporting network with the people keeping their eyes open for macro and micro anomalies in the natural environment and using simply-constructed earthquake survey facilities to issue earthquake predictions and warnings.[21] As late as May 1976, the city ordered further development of the system,[22] but there is no indication that the Tangshan earthquake was predicted with any accuracy through this system. When it struck, the city was soundly asleep. However, many other preparations outlined above were not taken in vain.

The Earthquake Strikes

The disaster that became known as the Great Tangshan Earthquake[23] began on 28 July at 03:42:53 AM while most of the city people were asleep.[24] People who were awake saw thousands of electric flashes in the eastern sky, like a red sun rising, and then heard a loud roar like thunder that lasted for some 30 seconds.[25] Then the earthquake hit. The strong tremors continued for 5–6 minutes. People were awoken in their beds startled and frightened, as one survivor remembers: 'The thunderous sounds of walls falling down mixed with the miserable cries of people, with the feeling of day never arriving and everything being annihilated.'[26]

When the earthquake struck, people reacted differently. Some of those who did not have time to run outside hid under their beds, large equipment, or in bathrooms and kitchens that often withstood tremors better than ordinary living quarters or factory halls.[27] Those managing to flee outdoors rushed to the streets 'not knowing whether it was the ground or their feet shaking'.[28] A soldier ordered to participate in the rescue operation described the chaotic scene of arriving at the Heping District a few hours after the earthquake had struck: 'I could see very many collapsed buildings, smoke stacks cut across in the middle, the deformed surface of earth, damaged roads, rescue cars sirens wailing endlessly, streets full of city people looking hopefully at us.'[29]

A relatively shallow epicentre (only 11 km underground) made the earthquake a category XI – the second strongest on the Chinese scale. Although located some 100 km away from the epicentre, different regions of Tianjin suffered VI–IX category damage depending on their geological conditions. As is usual in major earthquakes, several aftershocks also rocked the city. At 6:45 PM, a M 7.1 aftershock hit Ninghe County practically levelling it.[30] Up to the morning of 29 September, a total of 178 aftershocks of over M 4 were reported.[31] These made emergency rescue efforts hazardous and added to casualties and damage. Here the whole episode is dealt with as one single earthquake, because it would be virtually impossible to separate these earthquakes from each other.

Damage to the city infrastructure was daunting: two-thirds of buildings in the city were damaged to some degree, with only little differences between urban and suburban areas.[32] In the city, the Heping and Hexi districts were hit most severely, while in the rural areas Ninghe County, Tanggu, and Hangu District, which were located closest to Tangshan, bore the brunt of the earthquake. In Ninghe County, Lutai village was totally destroyed – no building was left standing there.[33] However, this did not mean that vulnerabilities neatly followed the rural/urban division so important in the GLF famine and the 1963 flood. Thousands of urban residents died or were injured and urban district casualties were highest among workers. Although geological reasons played a part here too, this was mainly because workers' residential areas were densely built with many old and dilapidated houses crammed together housing the workforce of urban *danweis*. Both in urban and rural areas houses had been constructed without following seismic design standards, as there had not been any before 1976.[34] Many of these houses could not withstand the tremors and therefore became unsafe during the earthquake, making workers vulnerable to this type of disaster.[35]

Apart from housing, the rest of the city's infrastructure also suffered tremendously. Roads were blocked or broken up, most phone lines severed, electricity generation failed and power was cut off, running water was cut off, gas lines were broken, and the loss of merchandise and medical equipment due to collapsing warehouses was huge.[36] Serious secondary damage also ensued. After the earthquake, 14 major fires broke out in the city, and seven cases of toxic gases being released into the air occurred. Several chemical fires ignited from broken vessels in chemical research institutes, factories, pipelines and medical raw-material warehouses. Conventional fires also broke out in home stoves, especially in the countryside, where many houses burned with their residents trapped inside. In cities, many restaurants shared the same fate when their hot cooking oils ignited. Collapsed buildings could also catch fire from live electrical cables, and a number of transmitter substations did the same adding to the general blackout. The weather was rainy on 28 September,[37] which made rescue more difficult, but may have helped in dousing the fires. Toxic leakages also occurred: a chlorine leakage at a factory threatened the lives of people downwind, but was contained and only three persons died and 18 were injured as a consequence.[38]

In human terms, the disaster also took a serious toll. According to the RERWHQ statistics, between 28 July and 12 August, a total of 23,938 people died, 32,572 were seriously injured and 83,827 slightly injured.[39] In the city proper, some 6,012 people died, while in suburbs and counties, Ninghe County alone suffered 16,097 dead, telling a grim tale of the severity of its aftershocks.[40] In monetary terms, the earthquake was enormously costly. Due to the size of the city, the value of damage was actually greater in Tianjin than in Tangshan. According to official estimates, the damage to Tianjin rose to 7.5 billion *yuan*. This consisted of 3.92 billion in direct costs caused to destroyed and damaged buildings, equipment, personal property of citizens, as well as losses to agriculture and infrastructure. In indirect costs, the city lost 3.581 billion *yuan* in 1976–9 in lost production.[41]

Emergency Response

For a short while the city and countryside were in disorder. Even a semi-official account admits that the city was affected by a temporary loss of control and many chaotic things were witnessed as people struggled through the streets seeking refuge from dangerous buildings and aftershocks.[42] The blackout, rain and loss of communications contributed to the situation. Many survivors had lost their daily necessities and family members, friends and neighbours in the ruins. Some did not even have any clothes on because they had been sleeping naked due to the summer heat. Stunned and frightened people filled the streets in the city and villages. Cries for help and of agony would come from ruins, where thousands of people lay buried under rubble, and from the streets, where people injured from falling debris mingled with luckier survivors.[43] An early estimate by the RERWHQ put the number of people forced to live out in the open in the city and countryside after the earthquake at 3 million.[44]

However, even though things were chaotic for a while due to the massive and unexpected destruction that had struck the city, the city leaders and residents were not totally paralyzed. Leading cadres reported to their post to the best of their ability shortly after the earthquake, navigating through the chaotic city scene.

As a cadre in the City Party Committee and Revolutionary Committee Office, who made the dangerous trip through the rubble from his home to City Party Committee headquarters right after the first earthquake, recounted: 'The roof and one wall of the building opposite [the City Party Committee building] toppled down crashing to the ground, making a deafening sound and filling the earth with dust. If I had gone forward just a few metres more, I would have been hurt or added to the list of fatalities.'[45]

All City Party Committee Standing Committee members are said to have arrived at the City Party headquarters within an hour of the earthquake,[46] and the first emergency order from the City Party Committee was issued already at 4:19 AM, some 30 minutes after the earthquake hit. As civilian communications were cut off, underground civil defence cables were used instead and connected to the City Party Committee switchboard. District and counties were ordered to conduct an immediate inspection of damage, injuries and fires and report them to the City Party Committee. Army radios were used in communications to more remote districts and counties, such as Ninghe and Hangu. In some counties, motorcycle couriers had to be used to establish communications.[47]

The City Party Committee held several emergency meetings in order to organize rescue and relief work. At its earliest meeting on 28 August, the CEO leader Lin Qiyu reported to the city leadership on the situation and was ordered to keep watch for aftershocks. Other than that, a full city cadre and personnel mobilization was ordered. Party and Youth League members, cadres and the militia were to be the core of earthquake emergency work. It was also decided to evacuate the population, especially the old, infirm and children, to safer locations and mobilize all medical resources in the city. Finance and Commerce Bureaus had to organize companies to secure the supply of basic necessities, and residents had to raise their vigilance to be on guard against fires, robbery and looting and maintain public order as well as against 'enemy activities'. Further, it was already now ordered that whenever possible, production would resume. The city leadership also divided leadership responsibilities between rescue and relief work, medical and first aid, maintaining public order, the provision of supplies, relocating the population, and resuming production among respective bureaus.[48] After that, Party Secretary Xie, vice-party secretaries and other leading cadres went to inspect the damage and lead rescue and relief work in person around the city.[49]

The City Party Committee convened another emergency meeting in the afternoon of the same day, further instructing all district committees to carry out the evacuation and relocation of the population, as well as supplying provisions for the evacuees. It was ordered that all dangerous buildings be immediately pulled down to minimize secondary casualties from aftershocks. Burying the dead had to be taken care of promptly. If cremation was not possible, temporary graves had to do. During the two days after the earthquake, the City Party Committee convened several more emergency meetings, deciding for example on constructing temporary shelters for earthquake victims and how to take care of their food supplies.[50] The City Party Committee meetings would continue to be convened throughout the emergency mobilization phase.[51]

The City Party Committee assumed leadership of the rescue and relief effort immediately after the earthquake and this demanded so much of their time and energy that it took until the morning of 29 August for city leaders to organize the actual City Headquarters for Resisting the Earthquake and Relief Work (for clarity's sake also abbreviated here as RERWHQ). It had 11 offices, designated to organize and lead the rescue and relief effort, including the relocation of people, and propaganda.[52] District, county and work unit party cells followed the city example and organized their local headquarters. In Ninghe County, the local Party Secretary had reportedly been buried under a collapsed building, but had been rescued. He then took command of the rescue operation ignoring his injuries, convening a County Party Committee emergency meeting and later organizing the local RERWHQ. As instructed by the City Party Committee, the county RERWHQ assigned leading cadres to go and inspect damaged areas *in situ* and lead rescue work there. The county Party Secretary was later praised for his leadership. Similar measures, establishing local HQs and dispatching leading cadres to the field to lead rescue operations there were taken in other districts and counties.[53]

At grassroots level, the streets and residential committees in city districts and communes and brigades in the countryside assumed rescue and relief effort leadership after the earthquake. They set up their own RERWHQs and organized rescue teams of residents.[54] The 66th Army stationed in Tianjin set up its own RERWHQ and soldiers engaged actively in rescue and relief efforts city wide.[55] Nevertheless, much of the immediate rescue efforts and the erection of temporary shelters were carried out spontaneously by the people on their own without much leadership. As one survivor described it: 'People could not find any tools, but time was life, so they competed with time, using their two hands to remove the dirt over injured people, and with their broken fingernails bleeding they persisted in their task. Afterwards, some of the rescued people would compare themselves to people who had "arisen from the dust".'[56]

Although propaganda would later take up people helping each other as representatives of correct socialist consciousness and revolutionary heroism, and thus being led by the Party at least in spirit, it is likely that many of these people were motivated by rescuing their loved ones, property and daily necessities from ruins and getting shelter from rain, not by some socialist ethos. Memoirs point to the fact that people tried to save their family members first and then ran to help their relatives or immediate neighbours and relied on mutual help to salvage daily necessities, furniture, clothing and foodstuffs from the rubble and dangerous locations during and after the immediate rescue phase.[57] That this was commonly the case was also indirectly shown by the fact that official propaganda attacked such behaviour by presenting individuals who saved public property instead of their family members as the true heroes of the night of the 28th (see below).

However, in many places individual party members did assume leadership of rescue efforts. Similarly, where they were able to function, street and residential committees would lead the rescue work, the transportation of the injured to hospitals, and clearing bodies.[58] Official and semi-official histories claim that the local Party assumed grassroots leadership effectively, but memoirs suggest that this was not very systematic, at least initially.[59] It is impossible to estimate exactly how quickly and to what extent officials were able to start directing activities during the first hours of the main earthquake, but it is clear that immediately after the earthquake people had to take care of themselves, and only after some while did the authorities' grip on the situation begin to improve.

The official propaganda machine was not damaged in the earthquake and, as during the GLF famine and in 1963, it began to serve the party line on the disaster right away. The *Tianjin ribao* came out the same day and began to bombard the population with disaster rescue and relief propaganda very similar to 1963, but with a Cultural Revolutionary spin. The propaganda expected that individuals discard their immediate needs and serve the rescue effort selflessly. For example, the newspaper told how in Ninghe County, the party secretary and his wife had been buried in rubble when the earthquake hit, but were rescued. After the rescue, his first utterance had been: 'let's go to the office, rescuing the masses is an urgent matter!', not a worry over his family. The county Grain Bureau chief had a similar encounter, when he and a number of his family members were buried when their house collapsed. However, as propaganda put it, after getting out, he immediately rushed to the office to organize the rescue of his class brothers and sisters and did not stay behind to save his family. When someone came and told him that four of his family members had died, he was very sad, but saw that as his own family had now died, he had more time to rescue class brothers and sisters, and he did not go to see his dead relatives.[60]

A variation of this self-sacrificing, or even selfless, storyline was told about those cadres and workers who stayed at their posts, not returning home despite knowing that their homes had collapsed. For example, a Xinhua Book Store worker was reported to have stayed at her post, not thinking about her little daughter at home, but rushing to the book store to rescue school books, thereby saving a lot of government property.[61] Similar stories were told for example about street-housing management staff,[62] leading cadres in many villages,[63] post and telephone offices,[64] Dakang oilfields,[65] hospitals,[66] the Railroad Bureau[67] and various individual party members.[68]

In 1963, the typical way to represent the same idea had been to tell stories of how a worker or cadre refused to stay behind in the city. In 1976, the same narrative technique was used, but examples were much more demanding for ordinary citizens to follow. This time people were supposed to be prepared to leave their children and relatives buried in the rubble and rush to their posts to save government property. Producing watches or saving books was clearly regarded as more important than trying to rescue one's own family members. Memoirs point out, however, that in most cases this kind of behaviour did not take place in reality.

A major part of the rescue operation was carried out by the military, which was commanded to the rescue by the City Party Committee and Military Party Committees with only a short delay. Over 30,000 soldiers and 1,000 military vehicles participated in rescue efforts (this figure included Tangshan, where Tianjin city also sent notable relief aid), transporting rescue and medical teams, the injured and materials. Soldiers would go through ruins listening for the sounds of cries for help. If they heard any noise, they would start digging trying to find and rescue the survivors.[69] According to statistics, they were able to rescue 6,254 people from collapsed buildings or under collapsed structures, some soldiers fainting from exhaustion. Apart from saving the living, soldiers also dug out 6,974 dead bodies and buried 14,781. In Lutai township, corpses were lying in rows on roadsides and the army transported them in trucks away from the scene for deep burial.[70] In Hexi District, the dead were put in plastic bags and taken away in carts to peasant fields and buried there.[71]

Troops also carried out the evacuation of residents to safer zones and cleared roads from rubble.[72] The army sent its medical teams to counties, such as Tanggu and Hangu to establish emergency clinics there. In military hospitals, personnel could be collapsing from exhaustion, but kept working without rest for days to treat the injured.[73] Apart from regular troops, as during the flood, the city's 300,000-strong militia was also ordered out at 5 AM on disaster day. They were led by their own HQ and district HQs.[74] Police units were also mobilized for rescue work.[75]

As the only real professional rescue force in the city, fire brigades had a major role in rescue work. They had to respond to the most dangerous calls in a chaotic situation. With major and minor fires raging here and there and communication lines to city districts and *xitong* cut off, local fire brigade leaders had to prioritize their targets. In Hangu, a local fire brigade had to respond to reports of a burning chemical factory, a large burning electric power transformer, and several buildings collapsing and needing rescue teams. By dividing up its teams, Hangu Fire Brigade was able to put out a fire at a crucial transmitter supplying power to Hangu and Ninghe,[76] while also handling a fire at a chemical factory. The factory had four large chlorine tanks, which had ignited threatening eleven 19 m-tall polymerization cauldrons at the facilities. Had these exploded, an area with a radius of 2.5 km would have been obliterated. The 19th Firefighting Squadron of the Tianjin Fire Fighting Brigade rushed to the scene with cauldrons shaking back and forth, hot ethylene alkaline spraying and scalding people, bricks falling all over. Fire fighters were nevertheless able to locate the fire source and, by fighting for three hours, extinguish it.[77]

Other city bureaus also organized their own rescue teams. The Transport Bureau had its first rescue team ready at 4:07 AM – all together 400 cars with 700 men in seven rescue teams. They rescued and transported over 10,000 people over five days. At the same time they transported needed materials such as food to victims.[78] If cars could not move because of the rubble blocking streets, three-wheeled carts were used instead.[79]

Special attention was given to Ninghe County and Hangu District, which had been hit worse than other parts of the city. Ninghe County especially received much attention. On disaster day, the City Party Committee dispatched first-aid teams and ordered air drops of medical supplies and the transportation of foodstuffs to these disaster areas as well as carrying out the medical evacuation of injured people there.[80] Field hospitals were also set up by the army in the county as well as in Hangu and special medical units were dispatched from the city and later from sister provinces. What made rescue and relief operation difficult in Ninghe was that a Yiyun River bridge that connected the city to the county had collapsed in the earthquake.[81] A pontoon bridge constructed by army engineers was already in place by 29 July. The City Construction Bureau participated by sending 400 men to the site to repair the big bridge, which was done quickly.[82]

A crucial part of the rescue operation was emergency medical work. It was among the first things ordered by the City RERWHQ on the morning of the 28th when the City Health Bureau organized its personnel to deploy first aid to victims. Over 30,000 city medical personnel participated in this campaign. This force consisted of 11,000 barefoot doctors in villages, a 10,000-strong staff at hospitals and other medical units, 6,000 factory doctors, and over 3,000 street-level health care personnel. In addition, the army and other regions sent 1,192 medical staff to Tianjin.[83]

All city district health branches organized small first-aid teams and dispatched them to the disaster sites to treat minor injuries and the sick, propagate health information and conduct preventive disease work. In the countryside, barefoot doctors were the backbone of rural first-aid work.[84] The general principle in treatment was that the slightly injured were given bandages and medicine, while only the severe cases were admitted to hospitals.[85] As hospitals could not admit all the victims, ten temporary hospitals and nine field hospitals were erected in the city, increasing bed places by 5,576.[86]

The city and rural hospitals were overwhelmed with the flood of injured people. Emergency wards in hospitals were so filled with injured people that moving around in them became difficult. Doctors and even bystanders had to mend fractured limbs without anaesthesia because of the lack of personnel and medical supplies.[87] A nurse described the situation at Tianjin Medical University Hospital on the earthquake day as follows: ‘The streets next to the hospital were full of injured people. They were brought on people’s backs, by carrying them, by taking them by hand or by themselves. We erected a temporary reception post on the pavement in front of the polyclinic, where the walking wounded were bandaged. … When I entered the hospital I saw a tragic sight, the corridors were full of injured people, standing and lying there, some receiving first aid, some getting bandaged, some lying silent.’[88]

While the city proper was pressed hard to provide medical care for the injured, the situation was worse especially in Ninghe County and Hangu District. This was recognized by the city leadership, which dispatched 13 medical-aid teams with 647 people there, and also assisted Tangshan. Other provinces also responded by sending medical teams to Tianjin. Beijing and Inner Mongolia sent 450 men.[89] Along with medical personnel, medical equipment and medicine were also provided to the injured and sick by the city medicine stations, medical companies and medical supplies companies.[90]

In Ninghe County, which was most severely hit by a strong aftershock, almost everything was missing. There was a scarcity of trained medical personnel, equipment, facilities and medicine. The first medical units dispatched to Ninghe had little in the way of medical supplies and equipment with them. Disinfected operation rooms were lacking and only tents existed. Because of a lack of disinfectant, river water was salted and boiled instead. When there was no lighting, torchlights were used instead.[91] The severity of the situation in Ninghe was revealed by the fact that medical personnel treated 54,664 injured in the county, while the city hospitals treated 12,585 Ninghe people for injuries, and 5,448 Ninghe and 1,600 Hangu people were evacuated to other regions to be hospitalized.[92] All together 120,900 people were treated in the city for different kinds of injuries and ailments, of which 23,100 were serious.[93] Moreover, the army medical corps offered medical help to 114,800 people.[94]

On 4 August, the city leadership announced that rescuing people was basically finished.[95] No figures of the number of rescued or victims were released, although the fragmented pieces of news and the figures on the numbers rescued by the army suggested that several thousand had been dug out from ruins after the first day of the earthquake. Although The *Tianjin ribao* continued to run articles on heroic rescue efforts and rescuers on its pages even after this, the main thrust of the campaign was now turned to the relief effort and resuming production in the city.

Relief Effort

Simultaneously with the rescue effort rose the challenge of providing food, long-term medical care, shelter, clothing and daily necessities to earthquake victims. The City Party Committee laid down the general principles of relief and reconstruction work already in its early report to the Party Centre on 29 July. The same principle that regions which had not suffered, or suffered only lightly, from the disaster, should help areas that had suffered more severely was applied in 1963 and in 1976. A similar emphasis on self-effort in relief and reconstruction (*zili-gengsheng*) was also clear. It was declared that relying on one's own efforts was to be the basic principle in reconstruction.[96]

As before, propaganda work was also high on disaster management priorities in 1976. The first emergency meeting on 28 August already gave a priority to it. As one of its first orders, the City Party Committee issued a propaganda outline and ordered restoring regular radio broadcasts to keep spirits high and calm the population.[97] It was emphasized how propaganda was to instil the correct fighting spirit and morale into the masses.[98] The City Party Committee also reported to the Party Centre that its first priority – listed before any other tasks such as rescue work or relief aid – had been to undertake propaganda.[99]

The key element in the propaganda effort throughout the disaster rescue and stabilization phase was the consoling telegram that the Party Centre sent under the name of Chairman Mao on the 28th to all regions that had suffered from the earthquake. The telegram itself, printed in the *Tianjin ribao* on 29 July, was a rather terse note declaring that Great Chairman Mao and the Party Centre extended their warm condolences to the disaster victims. The people and Liberation Army personnel were instructed to throw themselves into the struggle against the earthquake under the party's leadership in the spirit of persisting in bitter revolutionary struggle, going all out to reconstruct through one's own efforts, developing production and rebuilding the country. Moreover, it was instructed that people from all walks of life should study Chairman Mao's important instructions, take class struggle as their key link, and deepen and develop the struggle against Deng Xiaoping's counterrevolutionary revisionism and rightist deviations.[100]

In order to stabilize the masses' sentiments and strengthen their willpower, the City Party Committee decided to propagate the telegram broadly. Not only was it reprinted in the *Tianjin ribao*, but was sent to *danweis*.[101] The City Party Committee and the Revolutionary Committee also published their own letter of consolation to the city residents emphasizing the revolutionary fighting spirit under the title of 'Fear not hardship, fear not death'.[102] Many other regions sent their condolences as well to Tianjin. Ninghe County alone received over 3,000 consolation letters.[103]

The central telegram was elevated to a role of talisman and described as having almost wondrous qualities in and through the propaganda work. Indeed, in the first week after the disaster, the number one news item in the *Tianjin ribao* was not the earthquake or how to carry out rescue and relief effort effectively, but the telegram itself. Readers were told how it was being received with dance and joy by the masses and how people were crying hot tears when they first heard of this show of concern by Chairman Mao.[104] An old peasant was quoted in the newspaper of having compared the old society with the new one: 'In the old society when an earthquake struck no one cared [about us]. Now we have suffered some damage, but the Party Centre has sent us the consoling telegram and a consoling team, today we receive support for our daily needs, the two societies are really like day and night!'[105]

Apart from the consoling telegram, the Party Centre indeed sent a consoling team to the earthquake zones. It was led by the new Premier and Party Vice Chairman Hua Guofeng and had 13 members in it, including the Tianjin City Party Secretary Xie Xuegong. The team was divided into sub-teams which went to inspect different places. On 4 August, Vice Premier Hua's entourage arrived in Tianjin after having visited Tangshan first. On 6 August, Hua participated in an enlarged City Party Committee meeting where propaganda work was also discussed.[106]

While propaganda tried to keep public spirits high, the city government faced more mundane challenges. An urgent matter was how to keep the city population fed under disaster conditions. Many would go hungry in the first days after the earthquake, as foodstuffs and stoves for cooking were buried and destroyed, and stores and grain stations were not open, and in many cases had collapsed with the rest of the buildings. People had to dig out food from ruins and clear or erect temporary stoves for cooking.[107] For example, in Hangu this was specifically ordered by the district's RERWHQ.[108] In villages, the order for rescue work was to put people first, then foodstuffs.[109] Some help was provided by the army cooking units, but they could not serve all localities.[110] Survivors would also remember how right after the earthquake people shared what food they had with other survivors, but later in temporary shelters this spirit disappeared. A survivor would attribute this to 'Communist spirit',[111] but such solidarity between survivors is commonplace after major disasters, and usually does not last for long.

In its first report to the State Council, city leaders characterized the food supply situation as 'difficult'. The two major flour factories in the city as well as a majority of the Commercial Bureau's units dealing with food distribution had stopped operations.[112] The duty to solve food supply problems was assigned to City Grain and Food Service Bureaus. The first decision was to emphasize deliveries of dry food, especially baked wheat pancakes and biscuits to victims.[113] Already before this decision was made, all restaurants and canteens that were still standing were ordered to start to make pancakes for people. For a while customers were not asked for rationing tickets or money because many had lost these in the ruins.[114]

Providing enough dry food for residents was made difficult by the fact that, apart from the flour factories mentioned above, the only biscuit factory in the city had also suffered damage and had to cease operations temporarily. Factory personnel carried out rush repairs of their equipment to make food provision possible. The city's flour factories resumed operation at least partly by 1 August and the biscuit factory was able to repair its equipment and increase its production by one-third by this time, it was reported.[115]

In the meanwhile, food reserves were used up and the army delivered aid from other regions to the city, partly by air.[116] Shandong sent flour, Shanxi the 'biggest and best' potatoes, Hunan and Hebei rice;[117] Shanghai sent medicine, medical instruments, vegetables and clothing and Qingdao tankers with fresh water. Beijing, although it had also been hit by the earthquake, sent medical supplies, transport groups with cars and tons of pickled vegetables.[118] In order to secure provisions, the City Vegetable Company dispatched its buying teams to surrounding agricultural units.[119] As a consequence of the help, food supplies to the city actually increased by 30 per cent from the norm in early August.[120]

For the distribution of grain and vegetables many *ad hoc* solutions had to be resorted to. The Grain Bureau organized cars and trolley cars to distribute grain, taking full propaganda benefit from sending its leading cadres out to do some distribution work personally.[121] The Bureau also made a decision that relocated people could buy their grain from the nearby residential district grain stations with an introduction letter from the street offices of relocation areas. In Hebei District the damage to commercial facilities was so big that the responsible grain station units had to construct makeshift shacks and stoves to continue their operation. At the same time, to solve the problem of fresh water, the City Fire Brigade used sprinkler cars to deliver water to people.[122] Clearing and repairing wells and pipelines was also a priority operation that was undertaken immediately and lasted until 15 August, when the fresh water problem was 'basically' solved for city residents.[123]

The situation was even more challenging in rural areas and Hangu District, where some 90 per cent of buildings had collapsed including the local food storehouses. It took until 10 August to resume normal operations of grain deliveries in Hangu. In Ninghe, an early emergency measure by the city was to send in 50,000 kg of biscuits, cakes and other foodstuffs.[124] In Ninghe, the emergency delivery of grain began already on the 28th and every person was given 1 *jin* of grain per day for free until 10 August, when some delivery charges applied until 1 September. Only after this was delivery organized based on normal rationing principles.[125] The levels of emergency grain to earthquake victims were therefore close to the levels the flood refugees received in 1963. The basic frugality of relief aid continued in 1976, even if rationing had to be otherwise temporarily loosened under the circumstances.

By 1 August, the City Party Committee could report to the Party Centre that grain stores had 'basically' resumed operation and were taking 'temporary measures' to provide food for residents. Other food service trade was also gradually resuming.[126] By the end of August, some 210 million kg of relief grain had been delivered in the entire city, which was 30 per cent more than usual grain consumption. This guaranteed that residents were fed.[127] However, another big problem was how to arrange temporary accommodation for hundreds of thousands of residents made homeless by the earthquake.

Handling the Internal Refugee Problem

The Tangshan earthquake made the city face what amounted to a massive internal refugee problem. Some 68 per cent of residential buildings were damaged, and 20 per cent of them had totally collapsed or were beyond repair (see Map 3.1).[128] Some 1 million city people had to live out in the open. As one of its first orders, the City Party Committee ordered cadres at all levels to take care of evacuation (or relocation) work. This meant moving people into wide streets, parks, schoolyards, sports grounds and any open spaces away from harm's way and from blocking streets from rescue and relief workers and traffic.[129] The City Party Committee meeting on 1 August also instructed that temporary shelters be located away from tall buildings, narrow alleyways, smokestacks and power transmitters, and that fire hazard be monitored.[130] Apart from temporary shelters, people were housed in schools, universities and various *danweis*' public facilities that were still standing. Even partially ruined buildings were used for this purpose.

In order to help in the construction of temporary shelters, the City Party Committee assigned the Industrial and Commerce Bureaus to transfer 90 tons of plastic sheets, 1.95 million straw bags, 92,000 reed mats, bamboo stalks and sticks, wooden poles, nails, iron wire, telephone line, torches and batteries to districts and counties.[131] Reed mats could be used to construct 1.3 million m^2 of canopy and were funded by the central government.[132] The general instruction about the temporary shelters was that they had to be 'simple and practical', while still habitable in the winter.[133] This already indicated that the City Party Committee did not expect a quick solution to what was becoming known as 'residential villages' (*jumin cun*) or 'residential points' (*jumindian*), euphemisms for evacuee camps that rose up everywhere in the city.

Counties also distributed building materials from their own reserves.[134] However, in the lower levels of administration the lack of reserve materials led to resorting to many *ad hoc* solutions. In a Ji County commune the immediate emergency supply of building materials for shelters was solved by collecting all

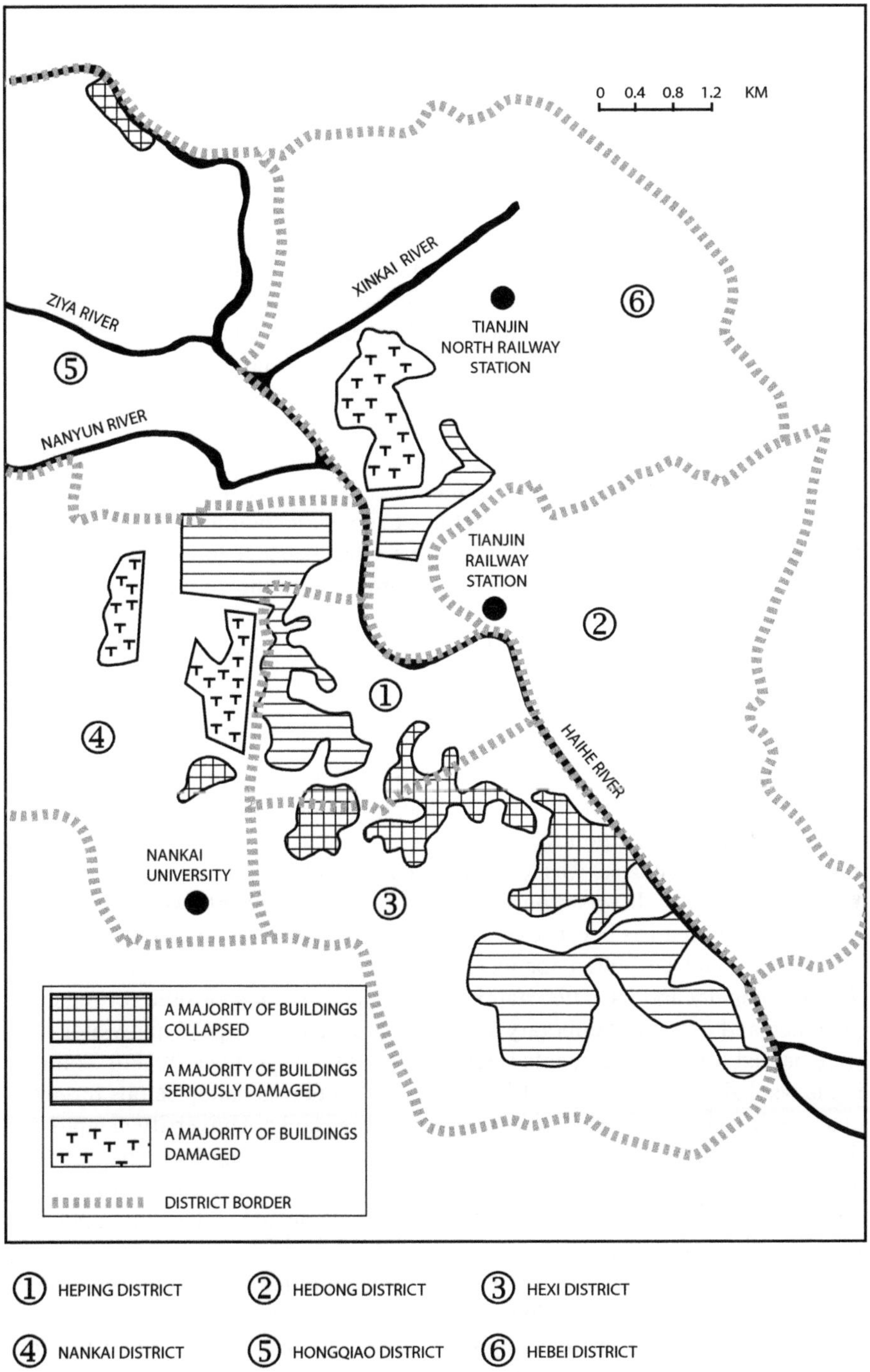

Map 3.1 Distribution of damage to buildings in the city areas in 1976

rain cloths, plastics and reed mats that could be found in the commune and using them to erect over 6,000 shacks.[135]

In the city, the responsibility of constructing and managing shelters fell to *danweis* and streets. As the City Party Committee ordered right after the earthquake, activists and masses were to be mobilized on the basis of constructing shelters in streets and open spaces.[136] The Construction Bureau also assigned material and manpower to the construction work. Moreover, army units erected 79,500 temporary shelters.[137] However, the masses were mainly to rely on their own strength in the process. Indeed, many people did not wait for orders or materials from above, but took matters in their own hands and constructed their shelters on their own and with the help of their neighbours, using whatever materials they could find in the ruins. Later, this caused city authorities headache when such self-help shelters blocked roads, overstocked and occupied water outlets, electricity, fire posts, public security sentry boxes and bus stations, etc.

The city left counties to handle the practical organization of shelter building largely by themselves. In Ninghe County three ways were used. One was that of using the Construction Bureau's workforce to construct buildings in a regular and planned manner. The second way was to let villages use their own workforce to construct buildings in their 'spare time' in order not to mess up their production tasks. The order was to construct buildings first for the relatives of revolutionary martyrs, the 'five guarantee' households, the poor and middle peasants, and only last to cadres and 'expelled' households, reflecting how also earthquake relief aid was expected to follow the pecking order of Maoist society. The third way was to let villagers rebuild their houses by their own initiative and cooperation.[138] The Ninghe mixed model was clearly used in the whole city with variations. It was largely based on the fact that people mostly helped themselves and the city authorities called this a policy.

One of the urgent tasks was to provide the new 'residential villages' in the city with running water. It was reported that in some cases people were getting sick in the evacuation areas because of a lack of clean water. As an emergency measure, water could be boiled for example in big kitchen cauldrons.[139] The City Water Company and its water stations in districts were responsible for repairing pipelines and drawing temporary lines to the evacuee camps. Such water lines were named 'making the people [feel] convenient – water lines'.[140] For example, in the Nankai and Hongqiao districts, seven different evacuation camps for earthquake refugees were located at such places as the Nankai University campus and the Water Park. They were provided running water by laying 1 km of temporary pipelines serving some 50,000 evacuees.[141] However, water supplies remained tight until the reconstruction of the city water pipeline and water station network in the early 1980s (see below).

As noted above, on 4 July the City Party Committee convened an expanded meeting of its standing committee with a consolation team from Beijing participating in the meeting. The meeting summarized the rescue and relief effort thus far and analysed the situation, declaring the masses' food, accommodation and daily necessities problems as 'preliminarily' solved. The Commercial Bureau was nevertheless ordered to speedily restore supply and adjust its methods to suit the new reality of 'residential points' and approaching winter. Furthermore, it was ordered that speculation had to be severely clamped down upon.[142]

With these instructions, the city bureaus responsible for the supply of food, daily necessities, and services had to adapt their service structures to fit the post-earthquake urban geography. For example, Heping District reorganized retail outlets for food, medicine and clothing to its relocation areas.[143] Adaptation also meant the creation of new supply policies and mobile units for the job. The *Tianjin ribao* reported about a supermarket whose staff organized mobile delivery units 'going among the masses' with soap, batteries, flash lights, pesticide, plastic sheets, rainwear, toilet paper, etc. This was called a 'consoling service'.[144] In the same spirit, the Coal Company reportedly began to sell coal in all quantities to shelters under the slogan of '5 jin is not too little, 100 jin is not too much'. Reorganization included units from medicine retailing, barbers shops and the City Gas Company.[145] In general, in most of the city districts the retail outlets could resume operation by the first week of August, at least partly. For example, in Hongqiao District almost half the retail outlets' staff were missing and one-third of them were still closed.[146]

Regardless of such attempts at 'consoling' the evacuees, their new housing arrangements were often very basic. They did provide protection from the elements (first it was rainy and hot, then the colder time of year began to approach), but they were uncomfortable both in the cold and heat.[147] In addition to discomfort, there was a constant fear of aftershocks, robbery (huts had no locks on doors if they had doors at all) and fires. As one survivor recounted it, living in self-constructed shelters right after the earthquake was miserable: 'It leaked when it was raining, roasted when it was sunny, there was no way to prepare meals, so you always felt hungry. You were terrified and enduring hardship there. But most frightening thing was the frequent aftershocks.'[148]

The temporary shelter could also easily catch fire as their materials were light and flammable, huts were constructed close to each other, their electrical insulation largely insufficient and fire protection equipment mostly missing or inappropriate. For example, at the Tianjin University compound 'residential village', two great fires took place destroying 280 shacks. In 1976–9 temporary shelters recorded all together 452 fires killing 52 and injuring 56 people.[149]

Many of the earthquake victims had lost everything in the ruins. Simply providing shelter, medical care and food for them was therefore not enough, but money and other daily necessities had to be provided for the victims as well. Up to the end of the year, the city helped 73,300 households with 286,600 people in them by providing 1.66 million *yuan* in aid. Suburban areas and counties were given substantially more, reflecting their higher damage rates. By the end of 1976 they were given 44.36 million *yuan*, of which 19.92 million *yuan* was for living expenses. A lot of materials were also donated to them for making winter clothing. The Civil Affairs Bureau gazetteer notes how in suburbs relief aid 'solved the food and clothing problem for 207,999 people'.[150] Padded clothing was provided for 63,100 people. Other regions sent Tianjin 75,800 pieces of clothing. In true

Maoist fashion, especially the relatives of revolutionary martyrs were given supplementary help in villages.[151]

While taking care of its own residents, Tianjin proper was also affected by a refugee problem, which nevertheless was much smaller than in 1963. Earthquake victims from Tangshan, Ninghe, Wuqing, Baodi, etc. and adjacent Hebei Province areas entered the city demanding medical care, nutrition and living quarters. As during the GLF famine and the 1963 flood, the City Party Committee ordered the Bureau of Civil Affairs[152] to set up a receiving and resettling organization for outside refugees under the RERWHQ. As in 1963, two *shourong* stations were set up in the western suburbs with canopy shacks and bedding for refugees. In addition, there were reception stations in every suburban district and county. Also as during the GLF famine and the 1963 flood, the policy was to send people as soon as possible back to the area of their *hukou* – to participate in rebuilding, as it was officially explained. Only those who could not be sent back right away were given temporary shelter.[153] This time there was no talk of using peasant households to accommodate refugees, but such a policy was obviously not feasible when the city itself had 1 million of its own residents living in evacuation camps and many villages of its own had suffered serious damage. Unlike earlier, this time the city itself was a disaster zone.

Refugees who had to be admitted to hospitals were given some pocket money and 1–1.4 *jin* food per day. When released from hospitals to be repatriated, the city also gave them some pocket money for expenses. Refugees who were not injured were given 5 *jiao* and 1 *jin* food per day. This money came partly from central relief funds. Repatriated people were paid their travel expenses and given pocket money of 1 *yuan* for the road for a day. To make sure that people actually returned to their homes as early as possible, rail and transport departments arranged special routes, cars and transportation. Reception centres handled over 10,000 people, 3,000 of whom were seriously injured. Centres also helped cadres, soldiers and workers who were in or assigned to Tianjin for some reason at the time of the earthquake and had suffered injuries or loss of property.[154]

Damage to Infrastructure

Much of the basic infrastructure of the city had been damaged in the earthquake and rebuilding it became one of the priorities. Roads and bridges had to be cleared of debris and repaired, and communications, electricity and water had to be restored. Clearing roads was a major concern. Not only had the earthquake thrown rubble and obstacles onto the roads, but thousands of temporary huts had been erected on the streets. In its meeting on 4 August, the City Party Committee decided to relocate people living on the streets to other places, clear rubbish and rubble, and increase traffic control.[155] This was a slow process, and the decision had to be repeated later on. Clearing rubbish from the streets was the target of a mass campaign. From the last week of July and early August, 2.5 million people participated in removing 320,000 tons of rubbish from city streets and public areas and 3 August was designated as a special health action day for the campaign. On that day, 800,000 people participated in clearing rubbish from the six main city thoroughfares. This action day was supported by the army, which sent cars and personnel to participate in the work. The city's trunk lines were cleared so that 62 main roads were passable by 8 August.[156] The Tianjin–Beijing railroad line was restored to operation on 7 August and the city's main bus lines were restored in 23 days.[157]

A number of main power lines had been severed along with many transformer substations that had burned down or were otherwise damaged. It took until mid-August to repair most of the power transmitters. In the meantime, dozens of small generators or small power stations were used to provide electricity to urban areas' key units. The priority was to restore electricity to the City Party Committee facilities, hospitals, railroads and the harbour.[158] Resuming communications was another immediate priority. This fell to the Post and Telecommunications Management Bureau (PTMB). Telephone lines were cut off totally in the city, but civil air defence underground lines were used for the rescue operation's communications. Civilian telephone lines were then restored gradually, first the main lines to the suburban districts and counties, then long-distance lines, and by the end of September internal telephone lines to 12,500 connections in the city were working again. In postal services it took three days to clear roads, to resume services towards Beijing. In the counties, such as Ninghe, radios had to be relied upon for communication with the city central administration.[159]

On 6 August, the City Party Committee could report to the State Council that, as of 4 August, city power transformers had been repaired. Trunk roads were clear for driving cars and the main communication lines had been restored.[160] Clearly this was too optimistic a report, as a number of power transformers were still down, clearing the rubble was still under way and telecommunications still mainly down, but the city leadership was probably under pressure to produce some good news for the Party Centre with Premier Hua Guofeng visiting the city.[161]

Repairing public utilities was the responsibility of the bureaus and companies that operated them. Right after the earthquake, the Public Utilities Bureau held its emergency meeting and ordered inspections of its facilities. Restoring the main fresh water pipelines was the primary task of the City Water Company and Water Engineering Company. The Electric Tram Company personnel had to repair electric cable lines and clear roads. The Gas Company repaired gas pipelines.[162] Companies drafted residents to help them in the work. For example, the City Engineering Bureau's Discharge Pipeline Office mobilized its own personnel and the masses to rush repairs of rainwater discharge lines. Several rainwater wells had collapsed or been buried under the ruins and discharge tubes were broken. This seriously affected some of main trunk lines by creating water logging. The office responded by bringing generators to pumping stations that had been cut off to resume power, and mobilized residents to clear rainwater wells street by street, well by well. Over 4,500 wells were cleared this way and the trunk line rainwater discharge problem was solved.[163]

Public Hygiene and Preventing Epidemics

When the earthquake hit, the weather was hot, humid and rainy. This created conditions conducive to the spreading of diseases. In the countryside, cases of dysentery, enteritis, secondary encephalitis (brain inflammation), etc. were reported.[164] Already the City Party Committee's second report to the Party Centre on 1 August noted that measures to prevent contagious disease from spreading were being taken.[165] On 4 August, the City Party Committee meeting decided that medical personnel had to establish a presence in the evacuation camps to carry out hygiene work there. General sanitation work, public toilets, clearing rubbish and exterminating pests had to be carried out promptly. The concentrations of injured and sick had to be constantly monitored. In the rural areas, drinking water and waste had to be properly handled in order to prevent diseases from spreading.[166] On 6 August, the City Party Committee held a telephone meeting with county party secretaries in order to organize anti-disease work by repeating these instructions.[167]

To carry out these decisions on public health work, mass mobilization of residents was used. The mass campaign to remove debris from city streets and public areas also targeted rubbish with the aim of removing and disinfecting the breeding ground for worms and mosquitoes, as well as increasing food and water safety, especially in the temporary shelter areas. Medical personnel acted as the core of this part of the campaign. The district had propaganda cars disseminating information on disease prevention.[168]

Health information was also disseminated in some articles in the *Tianjin ribao*, although the newspaper had otherwise very little such practical and useful information for earthquake victims.[169] On 6 August, it published 'Instructions by the City Hygiene and Disease Prevention Station'. The slogan was 'prevention first'. People everywhere were instructed to: 1) Construct temporary public toilets and assign people to take care of them. Defecating as one pleased was not allowed. 2) All garbage had to be collected in one spot in courtyards and cleared away every day. In suburbs garbage pits could be used for this purpose. 3) In the surroundings of temporary shelters, water pits and weeds had to be removed to prevent insects from breeding. 4) Harmful insects had to be exterminated using pesticides in potential breeding places every day. DDVP was recommended for this. 5) The Red Guards and Young Pioneers and the masses were to be organized to exterminate flies.[170]

Those street committees, which had done good work in carrying out hygiene work in the evacuation areas were commended in the press. In Nankai District a street Party Committee was commended for having organized doctors, barefoot doctors, health personnel and some rescue teams to form a health work group. Together with the masses, they had mobilized two quick campaigns repairing broken sewage pipelines and cleaning garbage dumps making many small ones into one big one and, with the help of a nearby work unit, using cars to haul garbage away. This had improved the hygiene level a lot. Reportedly, the garbage spot was emptied every day and hygiene inspections disseminated health information to people and conducted 'medical cleaning' in temporary shelters every two days.[171]

Rural areas also received attention. On a meeting on 7 August, the heads of the city's medical units decided that the three rural administrative levels of counties, *xiangs*, and villages (sic, but in 1976 this of course still meant communes, production brigades and teams) had to mobilize the masses for hygiene campaigns and strengthen education work on the subject. At the end of August, the city distributed anti-pestilence chemicals to counties, and the worst hit areas received DDVP and other pesticides together with 700 pieces of small portable spraying equipment. Apart from killing pests, 140,000 people were vaccinated in the worst hit rural areas. Dispatched medical teams were instrumental in this process and participated in disease prevention work actively through inspections, collecting disease information, and providing medical treatment. It was reported that because of the measures taken, the occurrence rate of disease during the high disease period in the spring of 1977 was actually lower than normal.[172]

An urgent matter in public hygiene was also how to dispose of the dead. One of the decisions of the emergency meetings on the first day was that burying the deceased must be taken care of promptly. If cremation were not possible, temporary graves would have to do.[173] In January 1977, Hangu District organized a special office to take care of the remains of the earthquake dead there. It was decided that in this district people buried closer than 200 metres from residential areas, wells or main roads were to be excavated and removed to an earthquake cemetery, which ultimately came to possess the remains of 950 earthquake victims.[174]

Maintaining Public Order

An immediate risk in a major disaster is losing public order. Tianjin was also affected by this, and maintaining public order was therefore high on the City Party Committee's agenda. The city leadership's first emergency meeting on the 28th already ordered militias, police and army personnel to patrol the streets and maintain public order.[175] In early August, The *Tianjin ribao* instructed people to be on alert for the activities of class enemies because allegedly 'small groups of class enemies will definitely use the opportunity of calamity to spread rumours and doubts in people's hearts, and engage in all sorts of ways in creating disturbances and bad activities'. The masses must not believe any rumours, but stamp them out, it was declared.[176] The City Party Committee meeting on 4 August decided that strict punishments were to be meted out to arsonists and thieves. Moreover, it was declared that the control of counterrevolutionary elements had to be strengthened.[177] This was probably due to the fact that some 'counterrevolutionary elements' spread rumours among the masses that the earthquake was caused by socialism and the CCP.[178]

There was an abundance of people patrolling. Street level RERWHQs organized small patrols with the participation of the local police station. These were responsible for street-level public order and safety work, political education, reporting on hazards, and surveying and striking down 'enemy activity'. At the same time, City Militia HQ also ordered that patrolling in small teams had to continue, while in *danweis* militias could participate in production, but also had to be on alert waiting for orders.[179] Propaganda also tried to connect the 'evil Dengist influences' to the public order disturbances after the earthquake.[180]

An early report to the Party Centre claimed that public order was 'very good' in the city, even if people were living outdoors.[181] Officially, the crime rate did not rise notably after the disaster. Some cases of hooliganism and robbery nevertheless took place. In Hangu, a mob looted a public granary on the night of 28–29 July stealing almost 40 tons of grain.[182] This suggests that at least early deliveries of emergency grain were not as efficient as had been declared, or that some people just took advantage of the situation. In Ninghe County, the earthquake sparked what was termed as 'feudal practices of witchcraft and sorcery'. Reported cases included such labels as 'stirring up trouble to serve one's own ends', robberies, 'conducting evil activities openly', exploiting another's plight by demanding sexual services from women, hooliganism, cutting down trees without a permit, damaging flood embankments and even murder. The occurrence of such bad activities rose by one-third from 1975 in the county. The police investigated the cases and adopted an attitude of 'severe punishments' towards criminals. All together 38 people were arrested and one executed in Ninghe.[183]

Politics of Disaster

A special City Party Committee meeting on 6 August declared that a great preliminary victory in the struggle had already been achieved, calling the situation in the city 'very good'.[184] This might not have been the first impression of an outsider at the time when large parts of the city were lying in ruins, some 1 million people were living in temporary shelters and the city was struggling to restore even its basic infrastructure. Nevertheless, the 'victory' was explained as being due to the people who were equipped with Marxism-Leninism and Mao Zedong thought, had been tempered by the CR and the campaign against Lin Biao and Confucius, and thus had elevated their political consciousness. The meeting also emphasized the need to deepen the campaign to criticize Deng Xiaoping and the struggle against rightist deviations.[185] After this meeting, and especially after the 8 August editorial in the *Renmin ribao* emphasizing the criticism of Deng Xiaoping as the key feature in the anti-earthquake campaign, the contents of earthquake propaganda changed notably. The anti-Deng campaign, which had been mentioned in the consolation telegram by the Party Centre, but had not yet been emphasized, now became the key theme in propaganda, as did the promotion of the 'new-born socialist things' brought about by the CR. The propaganda campaign also began to influence the conduct of relief efforts in an adverse manner.

When the Tangshan earthquake struck, Chinese national politics were going through the so-called Criticize Deng and Lin Biao campaign. It had begun in April 1976, when popular demonstrations in Tiananmen Square had led to the removal of Vice-Premier Deng Xiaoping from office. The campaign was mandated by Chairman Mao and conducted by a 'Leftist' factionalist grouping in the party leadership, which favoured the policies implemented after the CR proper (1966–9), and which was in the control of the propaganda system.[186] The earthquake was no reason for relaxing the campaign; instead, it became an occasion to intensify the rhetorical barrage on the 'Deng Xiaoping revisionist line' even further. The campaign showed once more how disaster management could become a victim of politics in Maoist China.

After the *Tianjin ribao* ran an editorial on 11 August on deepening the campaign against Deng and for disaster relief,[187] criticizing Deng became a daily topic in the newspaper's pages. The same article explained what the ideological significance of the campaign was about. According to it:

> In a society where social classes still exist, a struggle against the nature is closely linked to class struggle. History after liberation shows how every time a major natural disaster has occurred, also the struggle of two classes, two roads, two lines has got fiercer. The opportunists in the Party always try to use the situation of temporary hardship to twist the course of the revolution and restore capitalism.[188]

As it was now claimed, the choice in the relief and reconstruction work was therefore one between taking class struggle as the key and persisting in socialism or abandoning it and letting capitalism roam freely. In this principal question, Mao and Deng were said to hold opposite positions.[189] In another article, the newspaper put this choice as being between two lines: either making one's own destiny or apathy; relying on the masses or on a few men and material incentives; unity or individualism.[190] Propaganda even claimed that in 1959–61 the revisionist Liu Shaoqi – Deng Xiaoping clique sought to use natural disasters to stage capitalist restoration and this was also now the case,[191] which was of course a reference to the GLF famine and the pragmatic policies that had helped in ending it.

After this shift in the focus of the relief and reconstruction propaganda campaign, the *Tianjin ribao* gave a number of examples of how the class struggle manifested itself in practice in relief work. This included the question of arranging political night schools during the disaster situation, receiving relief aid, and individual efforts versus collective ones. It also tried to show how criticizing Deng had made recovery faster and motivated people to achieve even better results than before.

Units that had persisted in holding political night schools after the earthquake were now praised. This practice was explained as having helped turn the Criticize Deng campaign into a strength in reconstruction work.[192] In the countryside, peasants were told to study Dazhai and criticize Deng and thus gain strength in their struggle. Articles could tell how some people, including some cadres, thought that after the earthquake people should have concentrated all energy on repairing damage and resuming production, not spending their time at night schools. Fortunately, politically aware cadres knew better, and were able to convince the sceptics of the need for political study.[193] It was also reported that political night schools even helped peasants to increase their vegetable deliveries to the city.[194]

One reason for holding political night schools was given as the need to make people relinquish their superstitious belief in the Mandate of Heaven. In a story of a production brigade in the Western Suburban District, some cadres and members of the masses were said to have developed the symptoms of believing in the Mandate of Heaven and falling apathetic because of this. Local party leaders decided that such thinking could have serious consequences, and began to propagate the notion of man as the master of his own fate and strengthened political education through political night schools, where the central consolation telegram was studied and the situation was compared to the old society. This reportedly solved the problem.[195]

While holding political night schools to study the consolation telegram from the Party Centre was probably only a waste of peoples' time and energy amidst the rescue and reconstruction work, other aspects in this leftist ideological salvo were more worrisome. The question of the correct reception of outside aid was another issue that rose from framing the earthquake relief and reconstruction effort in terms of class/line struggle. This was illustrated for example in a case of a Hangu production brigade where, it was reported, a minority had requested higher levels to send them building materials and help in the reconstruction of their houses, but a majority had realized that they could not just 'raise their hands and wait for outside help'. The local party branch had realized that this was not only a question of how to construct shelters, but also what kind of spirit the masses should assume. *Zili-gengsheng* and revolutionary sprit had to be raised. Stronger ideological work therefore ensued in the brigade instead of asking for more relief aid.[196]

As such, making people help themselves was the strength of the Maoist disaster management style. But the question now arose of whether one should ask and/or accept relief aid *at all*. The *Tianjin ribao* would praise the cases where basic units had actually refused forthcoming relief aid, as in the case of a Hangu fishery production brigade, where the county had sent 250 reed mats as relief aid. The masses and cadres had reportedly refused the mats saying that they should go to areas, which were more severely damaged. Instead, the village reportedly erected 47 shelters completely from its own materials and work.[197] In another such story, one production brigade had suffered relatively heavily from collapsed buildings, but when the county cadres came to visit them their party secretary had proclaimed: 'Tell the county party committee that we will not cause trouble to the state, we can solve our problems on our own well enough.'[198]

In another story, Hangu Transportation Company cadres were reported to have decided not to ask the state for reed mats, or any other material help.[199] The people in Hangu were also quoted as saying:

> The relief aid from the state shows the superiority of the socialist system. But we cannot lie on the ground; stretch our arms upwards with our mouths open. A great earthquake means that we have to study *Dazhai*. Even in bad years we need to contribute to the state, recapture the ground that has been lost to the interference and damage of the revisionist and counterrevolutionary line of Deng Xiaoping and the natural disaster![200]

Whether to receive government aid or develop self-reliance was therefore framed as a line question and Deng Xiaoping compared to a natural disaster.

The same spirit applied to city districts as well. In Hongqiao, people were told to rely on their own strength and materials and not expect any help from the state in building temporary shelters.[201] In industry, a tanning factory was reported to have faced a similar predicament of either waiting for someone to come to repair facilities and equipment, or do it itself. This was, of course, not a practical question of skills, manpower, equipment and materials as it might have appeared to some people in the factory, but 'a struggle between two ideologies and political lines', the *Tianjin ribao* declared. In the factory, workers had therefore assumed the line of Mao Zedong thought, and tanners had become bricklayers and repaired their factory and resumed production on their own. Here a minority would have waited for help, but the majority did not.[202]

Framed as a line question, receiving relief aid became a test of the ideological standing of entitled recipients. Although viewing things positively, its meaning could have been just to encourage those units that knew they were capable of conducting repairs and reconstruction by themselves to do so and not burden the state, but in the highly charged Cultural Revolutionary ideological atmosphere the implications were much more severe. The logic was as follows: as the Dengists did not believe in the strength of the masses, expecting state relief aid and not relying on one's own strength was clearly Dengist. It opened the possibility that, come the next political campaign, a local cadre or common person who had accepted relief aid could be charged with an error of line. Therefore, it is possible that cadres who wanted to demonstrate their ideological purity over the real needs of their communities refused aid that their units would have been entitled to. This is even more paradoxical, when at the same time the Party was praising itself for its willingness to send in relief aid and the city leaders were applying for relief aid from the central government at the very same moment they were telling basic units to refuse it.

Whether this was a sign of a line struggle in the city leadership itself is uncertain. No cadre was attacked by name in the *Tianjin ribao* and later accounts written in the 1990s would have probably pointed to such struggle as leftist fallacy if it had taken place. However, Party Secretary Xie was a staunch Maoist as shown by his ability to stay in his position throughout the CR. He was nevertheless removed from his position two years after the earthquake when the Party was undergoing a cleansing of cadres too close to the 'Gang of Four'. Later in the 1980s, Xie did even lose his party membership. This may indicate that his actions after the earthquake did not give him much merit in the Party.

Attacking Deng Xiaoping was accompanied by the promotion of a number of Cultural Revolutionary institutions – the so-called new-born socialist things. These included barefoot doctors, sending educated urban youth out to the countryside, promoting young 'revolutionary' (i.e. Maoist) cadres in the Party, breaking down the barriers between manual and mental work and between offices and the real world, the Red Guard organization, the Dazhai model in agriculture, 'Sunny Courtyards' (*xiangyangyuan*) street organizations, model theatre (*yangbanxi*), and of course the role of Chairman Mao thought in guiding all social activities. On top of these came the nearly all-pervading rhetoric of class struggle. A good example of this was given in a rare article that had some practical value for its readers in the *Tianjin ribao* about how to use liquefied gas safely in temporary shelters. Even such practical themes were subsumed under an ideological barrage:

> Question: After the great earthquake in our city, a great many of the masses are living on the streets. Many of them have brought their liquefied gas stoves to be used outside, what should be taken into consideration here?
>
> Answer: At the moment the masses in our city are under the leadership of the party organizations, taking class struggle as the key link, enthusiastically engaged in the anti-earthquake movement. When using liquefied gas stoves on the streets, one must strengthen management and security, prevent class enemies from breaking [social order] and sowing chaos or losing the equipment.[203]

The 'new-born socialist things' were praised as the basis of the successful rescue and relief effort. It was described how barefoot doctors in villages had shown their proletarian feelings in the rescue of their class brothers and sisters of poor and middle peasants.[204] Members of the sent-down educated youth also reportedly began immediately to rescue people from collapsed houses in the countryside, digging members of poor and lower middle peasants out of the rubble with their bare hands.[205] This could be used to argue how the educated youth in the villages persisted in their determination to 'grow roots in the countryside', so the policy of sending educated youth to the countryside was therefore correct.[206] The role of younger, and therefore presumably more revolutionary, cadres was also dealt with in a number of articles that praised their achievements in rescue operations.[207] Otherwise, peasants were told to study the example of Dazhai in handling natural disasters.[208]

Red Guards and Little Red Pioneers were also organized for rescue and relief work. Red Guard members were reported to have stayed in their posts saving equipment at an earthquake forecast station, conducting emergency rescues despite their own injuries, taking part voluntarily in making medical supplies, and propagating Mao Zedong thought.[209] In the Maoist spirit, the masses were also told that it was absolutely wrong to rely only on some experts of special bureaus in reconstruction work.[210] Instead, it could be reported how 'local' (*tu*) and simple (*ben*) methods were used by the masses to repair pipelines, etc.[211] Theatre groups went to the field and performed model theatre plays to the masses, reportedly raising the revolutionary sprit of the masses greatly.[212]

As in 1963, the comparison with the old society and its inability to respond properly to natural disasters was a recurrent theme in propping up the legitimacy of the Party. People were to be told by old workers and/or peasants how in the 1939 flood the officials would use the situation for exploitation and speculation and leave the poor to die. In contrast, as one such article ran, 'Today, when Tianjin is struck by an earthquake, prices are stable, life is guaranteed' and Chairman Mao and the Party Centre show their deep concern for the victims, demonstrating the superiority of socialism.[213]

However, comparison with the past disasters had a distinct omission in it. In 1976, the 1963 flood was almost never referred to, and only the 1939 flood was used in comparisons. It is easy to see that this had all to do with the Leftist version of the Great Leap Forward and its aftermath, according to which at that time the revisionist line led by Liu Shaoqi and Deng Xiaoping tried to do away with the correct line of Mao Zedong. More specifically, it was about Party Secretary Wan Xiaotang and his fate as a 'counterrevolutionary' at the beginning of the CR. The city had been saved under his command using very Maoist mass line techniques that reputedly his kind of 'counterrevolutionaries' were against. Commemorating a deed by an alleged counterrevolutionary was out of the question in 1976. Pointing to the events of 1963 in 1976 would have contradicted the whole anti-Deng campaign that was connected to the earthquake rescue and relief work.

Reconstruction

On 1 September, the Party held a huge gala in Beijing with 3,500 representatives from advanced *danweis* and exemplary people in the anti-earthquake campaign including a delegation from Tianjin. Hua Guofeng presided over the meeting and all members of the central leadership were present. Also Tianjin Party Secretary Xie was invited. Mao could not attend as he was lying on his deathbed at nearby Zhongnanhai. Hua Guofeng gave a speech declaring a great victory in the struggle.[214]

Next day Tianjin held its own meeting declaring a victory, too. Just as in 1963, the *Tianjin ribao* listed the names of advanced units and model people in its pages.[215] When Mao died on 9 September, the daily ceased reporting anything else but Mao-related articles for about two weeks. After this, disaster-related news continued to appear in the paper, but gradually less frequently than before. With the declaration of the victory and death of Mao, the high phase of the propaganda campaign over earthquake mobilization can be said to have come to an end, but the reconstruction of the city had barely started.

Resuming Production

By the end of October, more and more signs of the city returning to some kind of normality appeared. Hangu Harbour resumed its operations.[216] Schools and kindergartens opened in many places.[217] The city now had to face the decision on how to undertake actual reconstruction. The question was basically over whether the city should emphasize production or the housing needs of its residents. As it happened, the city leadership opted for the former at the expense of the latter. All in all, Tianjin went through three distinct phases of reconstruction. The first phase was the immediate emergency repairs and structural consolidation work, then began the restoration of repairable buildings (*huifu*), and finally large-scale urban reconstruction took place.[218] As shown below, the last phase especially involved heavy politicking and the involvement of central government.

The city's industrial infrastructure had not escaped the devastation that had struck residential buildings. Indeed, the earthquake brought Tianjin industries to a grinding halt. As an early estimate by the City Party Committee indicated, ten generators in three major power stations in the city had been damaged and had halted. All four steel factories in the city had suffered serious damage. For example, the factory hall in the Second Steel Machinery Workshop had collapsed and 40 pieces of machinery were buried under the rubble. Because electricity and water supplies had stopped, a majority of factories had stopped operating. It was estimated that only one-third of industrial facilities were capable of carrying on, and the rest needed repairs to resume production.[219] Another early report indicated that only 2 per cent of 896 industrial companies in the city had continued operation without any damage, one-third were severely damaged, and the rest were slightly damaged.[220]

Resuming production became *the* preoccupation of the city leadership after the early emergency measures had been carried out. This could be seen in the rebuilding expenditure. During the first half of reconstruction in 1976–9, state assistance for rebuilding Tianjin was 1.894 billion yuan. Two-thirds of this was used to restart industrial production and the rest went to rebuilding housing and relief aid in general.[221] An emergency meeting on 28 July had already declared that factories and companies had to resume production while repairing their own facilities.[222] To glorify the efforts of the masses to this end, the slogan '*Di da zhen, ren da gan*' (roughly, 'a great earthquake asks for great deeds from people') was used frequently.[223] It was reported how under this sprit workers could achieve miracles. For example, in the Tianjin Heavy Machinery Factory workers had had the choice between running away from a dangerous potential explosion that threatened to raze the whole factory to the ground, or to try and prevent it. After doing the latter, they then restarted the factory in just 14 days through their own efforts. Reportedly, this would have previously (i.e. before the CR) taken six months.[224] In reality, recovery was at best partial in industry.

Other slogans used frequently were 'people are masters of their own fate' (*rending-shengtian*) and of course *zili-gengsheng*, which was already emphasized in the consoling telegram from the Party Centre.[225] These slogans made a virtue out of necessity. As we have seen above, much of the immediate rescue and construction of temporary shelters was indeed carried out by city residents and peasants themselves. Of course, this slogan was not telling the whole truth in that the city leadership itself was asking for aid from central government. However, on a national level, this policy was followed as the PRC government had refused to accept any foreign relief aid. Apart from this, which was already a considerable drawback on its own in terms of relief aid and recovery, the other problem with the slogan was that it could also have negative ramifications for the relief and reconstruction efforts, because, as we have seen above, it made receiving relief aid an ideologically dubious activity.

On 4 August, the City Party Committee instructed the slightly damaged factories to carry out their repairs on their own, while the city's resources were concentrated on those factories, which were too seriously damaged for self-repair. The most important industries had to be helped to recover first. Food-processing factories were a special priority, as well as factories providing necessities for relief work. The aim was to resume production at 50 per cent of factories within three months and 80 per cent by the end of the year.[226]

A late August meeting of relevant bureau-level leaders on the problems of industry and transportation declared that the general situation in resuming production was 'very good' and that 905 factories had already resumed operation. That said, it was admitted that many of them had done so only in a temporary and partial manner. Moreover, most of the resumed factories were small. Only few of the large backbone factories were fully or even partially operating. Apart from damage, raw material shortages were a constant problem. The daily output of industries was estimated to be only one-tenth of that before the earthquake.[227]

To increase the speed of recovery, the meeting made a number of general decisions. Reliance on mass mobilization was evident once again. All companies were told to mobilize the masses fully and rely on their own strength to resume production. However, the City Party Committee had to admit that it could not bring about full recovery on its own. In repairing larger industrial facilities with serious damage, the city reported the problems to the central RERWHQ and asked the State Council and its respective ministries for appropriate support. City leaders were confident that they could rely on the masses to provide construction materials such as bricks, tiles, ash, sand and stone, but asked the central government for steel, timber, and especially cement, because they were essential for the reconstruction of factories and residential buildings and the city did not have enough of them. The city also needed coal.[228] As material shortages were great, old materials also had to be fully utilized in repairs and warehouses were to be cleaned out in search of anything that could be used in reconstruction, it was ordered.[229]

The sequence of recovery was deemed important, too. A report on the damage caused by the earthquake on 15 November described well the chain reaction that happened when the flow of raw materials was disturbed. In this case, the halting of the Number Three Cotton Yarn Factory had meant that one-quarter of the city's cloth weaving factories (65 units) had to stop operation. In the chemical industry, halting the Red Flag Chemical Plant that produced boron (B) affected 20 other units in glass, medicine and other industries.[230] The rigidity of the planned economy in finding alternative sources of raw materials for downstream factories was thus revealed. While good at emergency support, the Maoist command economy's ability to induce industrial recovery was less impressive.

The recovery of industries went on gradually during the autumn of 1976, but the work clearly suffered a further setback in the 15 November earthquake. By this time, about 90 per cent of the light and moderate damage and about half the serious damage to industrial facilities by the earlier earthquake had reportedly been repaired. The new earthquake brought damage equivalent to about half of these repairs, although some one-third of damaged structures had not been damaged before.[231] Therefore, repairs had to start all over again in many places, and city authorities had to acknowledge that they would not be able to fulfil the city production quota for the year.[232] City revenues from its industries went down by 600 million *yuan* in 1976.[233] Therefore, contrary to what the official propaganda line in the press led readers to understand, city production in 1976 shrank notably because of the earthquakes. The city leaders nevertheless pledged to the central government that they would accomplish reconstruction and fully resume production in three years. However, for this they now asked for support from the central government in a 'big way'.[234] As seen above, substantial aid was indeed forthcoming, but even with it and all the mass mobilization for self-repairs, the ambitious reconstruction targets were missed by a mile.

The countryside surrounding the city had also been severely hit, especially those areas located to the east of the city proper. In its meeting on 4 August, the City Party Committee instructed that despite the disaster, agriculture should aim for a bumper autumn harvest.[235] This was a tough call to meet. The tremors had lifted dust and caused salty water to rise in the fields. Irrigation equipment was damaged and even some minor flooding of fields and villages took place as river embankments collapsed in some places. In Ninghe County, dust covered and salt water rose over 40,000 *mu* of fields, and water could not be removed from 170,000 *mu* because of broken equipment. As a result, 98,000 *mu* of fields was laid desolate. In Hangu, one-third of fields were damaged because of dust, cracks in the earth, and salt. All water pumps, three-quarters of wells, and one-third of sluices on the fields were damaged, water channels collapsed and their embankments mudded the surroundings. Many farm animals also died when their sheds collapsed on them.[236]

As a solution, the City Party Committee ordered the strengthening of field management and anti-pestilence work in communes. The Industrial Bureau sent small teams to repair irrigation equipment, agricultural tools and electricity generators in rural areas. As in 1963, the city gave orders to resume collective sideline production in order to accumulate funds for reconstruction.[237] However, no individual household sideline production was mentioned and no particular campaign for sideline production was undertaken, as had been the case in 1963. Mobilizing the peasants for production campaigns was also used in the countryside. On 5 August, all districts organized a campaign to remove water and mud from the fields, replant vegetables and generally strengthen the autumn planting season management.[238]

In August, recovering agricultural production through mass mobilization began to be stressed even more. Many villages held their own 'deepen criticism of Deng, study Dazhai, seize a bumper autumn harvest' meetings to increase production of grain, cotton and sideline products.[239] Later that month, it was declared that a mass movement of 1 million peasants to promote agricultural production and field management had been raised. The slogan was now 'do not let production drop after a disaster, strive for a bumper harvest'.[240] The peasants were told to study the experience of Dazhai on natural disasters and how it showed that peasants should be putting production first and life [i.e. living conditions] second.[241] Another slogan reportedly used by poor and lower middle peasants was: 'the heavily hit brigades will not lose in production, the slightly hit will get a bumper harvest'.[242]

The emphasis on production over living conditions said much about the underlying disaster management priorities in the Maoist era: the 'cult of production'. It had been the main cause of the GLF famine, had strongly influenced the city leaders' decisions during the 1963 flood, and in 1976 both city and rural residents experienced this in their daily lives when most resources were directed to reconstruction of production facilities. As seen above, people were even told that expecting relief aid to improve living conditions was wrong and showed incorrect political awareness.

However, with all the losses to the city, some small groups actually benefitted directly from the earthquake. With all its praise for the policy in propaganda, the need for manpower in reconstruction made the city leaders decide to cancel the policy of the rustication of 'educated youth' from the city, which had begun as a measure to bring the Red Guards under control in 1967–8 and then continued as a 'new-born socialist thing'. Many recently graduated people who had been already assigned to be sent down to villages were now informed that they could stay in the city and became construction workers instead. Moreover, a number of already rusticated youth were allowed to return to Tianjin from as far as Heilongjiang to participate in reconstruction. They were coined 'earthquake stayers' (*zhenliu*).[243] The recall even included cadres in the May 7th Cadre Schools, a penal colony system established during the CR for the re-education of cadres through hard agricultural labour and brainwashing. These cadres were now told to return to the city to participate in the reconstruction.[244]

Rebuilding Residential Housing

The earthquake had destroyed or damaged the homes of over 1 million people in the city proper and therefore solving the housing problem became the hot issue for city leaders. At the 4 August meeting, the City Party Committee ordered damaged buildings to be inspected and those that could be repaired had to be.[245] Apart from this, it was decided to construct 10 million m^2 of temporary residential buildings, which was an upgraded version of temporary shelters. To achieve this target, the city leadership decided to mobilize all resources, using all expert units and materials to construct buildings on free plots in the city and its suburbs and at the same time provide companies and streets with materials for reconstruction on their own under the respective city districts' management.[246] However, as we have seen, resuming industrial production was given priority over residential reconstruction needs.

In a Maoist fashion, much of the initial reconstruction was conducted by ordinary city residents themselves drawing on local resources, i.e. using bricks and other material from collapsed buildings, and often helped by their relatives, neighbours or work units. On National Day, the City Party Committee demanded that the task of temporary housing be basically completed by the end of October. By the end of 1976, 123,371 buildings, 96,245 of which were in the countryside, were reported as constructed and by the beginning of 1977 it was declared that enough residential buildings had been erected, 'basically' solving the housing problem.[247] However, this was very basic indeed. As the City Party Committee report to the State Council on the smaller earthquake of 15 November indicated, damage caused this time was as big as what had just been repaired in the past three months, and the number of people who had lost their homes still exceeded 700,000.[248] Many people died in the November after-quake because they were living in hastily repaired houses in which, for example, the upper floors were in ruins while people occupied the basement floor. In many such buildings, the already weakened structures could not withstand the tremors in November.[249]

Therefore, solving the housing problem 'basically' meant only that all evacuees had some kind of roofing over their heads. A breakdown of the housing numbers of evacuees reveals that temporary shelters were constructed for 300,000 people. Classrooms in schools and universities were used to house 200,000 people; rush repairs of slightly damaged buildings solved the problem for 100,000; *danweis* housed 100,000 people in their premises; 100,000 old people and children were lodged with their relatives; and constructing temporary housing solved the problem for 300,000 people.[250] Therefore, by early 1977 everyone probably had some place to stay, but living in a newly constructed building was still a faraway dream for most of the evacuees.

The city authorities quickly realized that nothing is as permanent as temporary. The residential villages and temporary housing were there to stay at least for a while and therefore needed regulation. On 12 September, the police issued a statement emphasizing how residents' lives and security also had to be protected under temporary housing conditions. Several construction regulations for fire protection were issued and fire-fighting equipment had to be provided to the 'residential villagers'. Security patrols were organized based on militias, guards and public security personnel.[251]

As during the GLF famine, the city authorities also faced the challenge of individuals and *danweis* engaging in self-help in ways that messed up the city plans. This time it was about building temporary shelters and demolishing damaged buildings. In order to bring these activities under city control, the City Revolutionary Committee issued an order on 28 February 1977 on 'Strictly Controlling and Demolishing Earthquake-Damaged Buildings'. It was aimed at *danweis* and individuals who were demolishing buildings for their own 'chaotic' purposes. The order instructed that those buildings that could be repaired should be repaired and not demolished. Damaged apartment buildings in particular should primarily be fixed and consolidated and not brought down. The district or city authorities had to approve all demolition work. The spirit was to have unified management and follow the central instruction of 'primarily building, secondarily razing'. The city now defined repairing damaged houses as its key policy instead of rebuilding new ones. It was further decided that repairing and reconstruction had to start based on the criteria of buildings being able to withstand M 8 earthquakes.[252]

The city also ordered that all temporary buildings be registered and work units pay compensation for the public areas which their temporary shelters occupied. Structures that blocked roads had to be cleared. Furthermore, it was instructed that the shelters or materials used in them had to be managed properly, not turned into private property, sold or bought individually – even if many of them were constructed by people themselves. When people returned to permanent housing, temporary shelters would have to be removed.[253] The regulations indicate how unauthorized businesses with buildings and building materials had emerged under reconstruction conditions, where such commodities certainly had their value. Memoirs also point out that people indeed could trade their huts with each other.[254]

While repairing damaged houses was made the main policy, many buildings and blocks were beyond repair. The city now concentrated its rebuilding efforts on the complete rebuilding of selected residential areas. In late August 1976, the city leaders decided to concentrate city-level efforts on most severely hit zones in urban districts and rebuild them completely. Each of the six inner city districts selected one zone to be rebuilt. These construction sites became known as the 'old six' and included Heping District Guiyang Road, Hexi District Dayingmen, Nankai District Dongnanjiao, Hebei District Huangwei Road, Dongxi District Dazhiguhoutai, and Hongqiao District Dahutong. In 1977, the 'small five' districts were added to the reconstruction zones.[255]

The reconstruction of the 'old six' areas lasted until 1979 and the 'five small' until 1980. The projects achieved certain success. New well-lit modern apartment buildings and public parks had been constructed in areas which had been dominated by *hutongs* and dilapidated housing. In the new houses, people now had an average living space of 5.75 m^2 per resident, up from just 3.3 m^2. Industries and warehouses were relocated away from the rebuilt districts, making traffic and living there more comfortable. The downside of this approach was that the results could be enjoyed only by a fraction of earthquake victims, as only about 84,000 people were housed in them.[256]

The central government was not impressed by these achievements, but before it intervened in the reconstruction directly, it reorganized the city leadership. Party Secretary Xie was removed from office in May 1978 for 'serious mistakes during the Cultural Revolution'. The Party Centre first made Lin Hujia and then, in December, Chen Weida Tianjin First Party Secretary. In August 1979, Deng Xiaoping visited Tianjin stressing the need for further organizational rectification in the city, which meant in practice the cleansing of cadres who had risen to their positions during the CR and replacing them with new less leftist cadres. Tianjin city and party administrations were under rectification from 1978 June until the end of 1983.[257]

Attention was also directed at the reconstruction of the city when Vice-Premier, State Construction Commission Head and Central Committee member Gu Mu came to inspect the city at the end of 1979. The city still had 1.1 million people living in huts and temporarily repaired housing. The remnants and ruins of the disaster were still visible everywhere. Gu Mu expressed annoyance that the situation was still so bad in the city. He proposed that the city appeal to the central authorities to solve the Tianjin situation in a 'Tangshan manner', where central assistance had greatly contributed to the recovery of the city. Following this cue, the new Tianjin Party Secretary Chen Weida, went to Beijing with such an appeal. Not surprisingly, the central government was receptive. The State Council sent a team of cadres, headed by a vice-head of the State Construction Commission to inspect the situation in Tianjin and then issued a report on the reconstruction of residential buildings in May. According to it, the reconstruction process had emphasized the resumption of production too much and had thereby ignored the plight of residents.[258]

Apart from mentioning the housing problems of the population, the report also listed other shortcomings in reconstruction work. City water supplies were inadequate and uncomfortable, traffic was often congested, and there were not enough schools, hospitals or shops. The report also noted how the situation had affected 'stability and unity' in the city, which was the political slogan of the time after the *de facto* ousting of Hua Guofeng in late 1978 and the political struggle around the Third Plenum of the 11th Central Committee, which had decided to set China on reform footing. According to the report, problems had to be solved comprehensively and in a united and planned manner. The state therefore had to concentrate its financial and other resources on solving Tianjin housing and infrastructure problems, and the signs of the earthquake had to be cleared away from Tianjin in a 'not too long a time', it was declared.[259]

Tianjin leaders acted on the report and a new phase began in reconstruction work, this time with an emphasis on residential housings and infrastructure. City leaders turned the central report into a plan of action in August 1980, making a decision to demolish temporary shelters and speed up residential construction to house the people who were living in temporary housing. The city leaders established the Reconstruction Leading Small Group under new Mayor Hu Qili, and a respective Reconstruction Headquarters (RHQ), which was also headed by Hu. As usual for such a campaign, all districts and streets established their own RHQs. With the help of centrally dispatched experts, the city drafted a three-year reconstruction plan.[260]

The plan for 1981–3 was to build 6.5 million m^2 of new buildings, 4.2 million m^2 of which were residential houses. Of these, 3.7 million m^2 were to be constructed in the ten new residential districts, which were to be opened in the city outskirts. In the city, 43 smaller zones were to be rebuilt. The plan also included some 20 km of new roads and other infrastructure. Points of traffic congestion were to be cleared from a number of roads and the roads were to be widened. The whole city water-supply system with water stations and pipelines was to be redrawn and rebuilt.[261] New water pipelines covering 436 km were to be installed, mostly new high-pressure pipelines. Two new water stations went online improving water supplies notably. The overall arrangement of roads was improved with 14 new or wholly reconstructed major roads and six new bridges over Haihe. New shops, schools, kindergartens and clinics were opened, as were new police stations, public toilets and administrative buildings, and the number of beds in city hospitals increased. Eighteen new bus routes were opened. On average, the living area grew from 3.5 m^2 to 4.4 m^2 per resident.[262]

There are different statistics on how the second phase of reconstruction was financed, but it is clear that subsidies from the central government played a big role in the drive. According to the City Construction Bureau Gazetteer, 2.46 billion *yuan* was used in the reconstruction of housing in the city in 1981–4. Central government paid 320 million *yuan* a year in reconstruction assistance, and a further 200 million in other subsidies. 100 million *yuan* per year came from company profits, on which a 5 per cent surcharge was levied for reconstruction. Furthermore, *xitongs*, *danweis* and private people used 200 million *yuan* a year of their own funds on reconstruction.[263] Wang Yuezhen, a former Head of the City Finance Bureau Enterprise Section, argues that from 1976 to 1985 central government supported reconstruction with 4.679 billion *yuan*, or about 82 per cent, and only a fraction of the rest of the 5.679 billion *yuan* used in reconstruction came from the city budget (4 per cent) and companies themselves (14 per cent). From 1976 to 1980, the central government subsidy varied between 350 million and 380 million a year, but from 1981 to 1983 it rose on average to 830 million *yuan* a year, and remained at 720 million *yuan* in 1984 and 670 million *yuan* in 1985, making the intense reconstruction drive possible.[264] However, Wang's figures add up only if they also include the 20 per cent that the city and its units paid for the reconstruction.

Housing the last evacuees still living in temporary shelters created some political drama. In June 1981, Vice-Mayor Li Ruihuan announced in an all-city cadre meeting that temporary shelters were to be erased from the city landscape by the end of the year, so that the people living in shacks would be able 'to eat their New Year's *jiaozi* in new houses'. The houses for these people were to be 'borrowed' by the City Government for a year from *danweis* that had constructed their own housing, but had not yet allocated them or into which residents had not yet moved. The whole administration down to *danwei*- and street-level was made responsible for finding housing and coordinating the demolition of the temporary shelters.[265]

Related to this move, specific internal party orders ordered the inspection of cadre discipline in relation to housing. In October, a new and reformist Mayor Hu Qili stated in the City Government Work Report that leading cadres, and especially cadres in charge of house distribution work, and their relatives and friends whose houses were allocated or arranged by these cadres, would have to make way for the still remaining earthquake refugees who had not yet been granted housing. Cadres and their relatives who were already living in new houses would have to return to their old houses and let earthquake refugees live in the new ones. It is unclear how much of this actually happened, but in early 1982 temporary shelters were finally empty and razed to the ground.[266] Already by 4 August 1976, the City Party Committee meeting had emphasized that reconstructed buildings were meant for the masses, and that leading cadres must not claim privileges on them.[267] Clearly, not all leading cadres had followed this decision, but used reconstruction for their own benefit and, as during the GLF famine, acted selfishly.

In the end it was burying Mao and his emphasis on local self-sufficiency that made reconstruction a success in Tianjin. But this required massive central assistance and time. The idea that workers and residents could rebuild their factories and houses and resume production at the same time was a typical Maoist pipedream. The masses could do some things, and they were useful in rescue and emergency mobilization, but rebuilding an entire city was not the same thing as building a dyke or planting vegetables. Although mobilization-based disaster management had its merits, it also had its limits and the 1976 earthquake clearly demonstrated these.

The Legacies of the Earthquake

By the end of 1981, the shacks finally disappeared from the cityscape and were turned into 600 million kg of rubble. Some compensation was paid for self-constructed shacks, whose residents were now given new modern homes.[268] Yet, reconstruction continued. The City Construction Gazetteer notes that reconstruction was brought to an end in 1983,[269] but this means only the inner city projects. In 1982–7 17.12 million m^2 of floor space was constructed in the city – as much as during the previous 31 years.[270] The construction of ten new residential areas was expanded to 14 in suburbs surrounding the city. This lasted until 1990 and during it a total of 36.60 million m^2 of residential buildings were constructed. Indeed, reconstruction was officially fully over only in 1990 and in some places such as Ninghe County, even relief aid continued until 1989.[271] However, not all these building were related to the earthquake anymore, as the city was growing during this time in the general growth boom brought about by Deng Xiaoping's and Zhao Ziyang's economic reforms. Where the actual reconstruction phase ended is therefore a matter of conjecture rather than a specific point of time or project. Here it is placed in 1985, when major central assistance for reconstruction ended.

Like the flood in 1963, the earthquake left a legacy in the city. City planning and building regulations were revised based on the experiences gained from the disaster. In surveying the damage it was evident that older structures, such as small residential buildings in *hutongs*, had not withstood the tremors very well. Unsurprisingly, modern steel structures had fared much better. Furthermore, those industrial and residential structures that had been reinforced before the earthquake had withstood the tremors quite well. For example, two of the city power stations had been reinforced before the earthquake and no damage occurred to important areas in them.[272] Preparations had paid off in this respect, but they had been too small-scale in comparison to the rest of the city.

New national building standards were introduced due to the Tangshan earthquake and also in Tianjin new construction projects were to follow strict regulations on earthquake-proofness; new buildings had to be able to withstand earthquakes on the same magnitude as the city had just witnessed.[273] Attention was also paid to preventing secondary damage by urban planning. The essential public utilities (water, electricity, communications, traffic, gas, etc.) were to be deployed in network fashion, so that there would be multiple supply channels in case of emergency. Electric cables were put underground, key *danweis* were equipped with emergency power generators and multiple water sources were used in reconstructing pipelines. Attention was paid to constructing wider streets and increasing entry points into the city. The city's street network was reconstructed to allow better movement of vehicles.[274] In 1985 'Tianjin City Areas' Anti-Earthquake and Disaster Regulations' were issued and all city bureaus and *danweis* were instructed to plan earthquake protection and emergency work measures ahead.[275] Building new storage facilities and storing more emergency materials, equipment and foodstuff were also carried out.[276] As a Chinese account on the disaster puts it, now Tianjin should be disaster-ready.[277] Fortunately for the city, this has not been tested in practice, and hopefully will not be in the future either.

Not only did the earthquake leave its physical marks on the city, but also many residents would have to carry the scars of it for the rest of their lives. The disaster left over 500 widows/widowers without anyone to rely on, over 600 orphans and over 500 handicapped. Primarily these people were delegated to relatives and friends to take care of them, but the 'five guarantee' subsidy was also paid for them if these people had economic difficulties. In suburbs relatives who had children delegated to them were paid 7 *yuan* per month, in cities relatives received 10 *yuan* (for babies under six months old, this was 15 *yuan*). Children without relatives or friends to take care of them were taken care of by government orphanages.[278]

All together, 242 people were paralyzed in the earthquake. The city gave them 28.8 *yuan* per month for living expenses, but those who had suffered their injury in the line of duty (such as soldiers and others participating in rescue work) got a 5 *yuan* increase on this subsidy. For the peasants who were injured, the policy was primarily to rely on individuals and collectives, and secondarily on government subsidies. The levels of subsidies varied from 10–15 *yuan* per month, indicating once again the urban bias in Maoist social welfare practices. Households with injured people were also to receive 20 *jin* of grain, 2 *jin* of meat and 1 *jin* of sugar per month as a subsidy. Production brigades had the duty to take care of the living expenses of orphans, but the government could, if needed, subsidize them with 9 *yuan* per month.[279]

In Ninghe County, the state established a relief centre for earthquake orphans and elderly people with no relatives to take care of them because of the earthquake.[280] In 1981, the city gave Hangu District 400,000 *yuan* to construct a clinic for treating earthquake victims.[281] In the same year, the city also built a clinic in Tangu for paralyzed people and others who could not take care of themselves anymore.[282] It is clear that only a trifling part of relief and reconstruction funds went to helping disaster victims this way.

On the tenth anniversary of the earthquake in 1986, the earthquake victims' campaign was finally granted a lasting reminder in the centre of the city when a memorial monument in Huangjia Park was opened. The memorial was inaugurated by Vice-Premier Wan Li and had an inscription by the then Party Secretary Hu Yaobang. It was, and still is, located in a busy street corner in downtown Tianjin, reminding passers-by of Tianjin's difficult past with natural disasters.

Conclusion – Comparing the Three Disasters

What the Left was attacking in 1976 was ultimately a straw man created for propaganda purposes. As the 1963 flood showed, 'Dengist' disaster-relief policies and practices did not differ notably from what was carried out in 1976. When stripped of rhetoric and the differences in the physical nature of the three different kinds of disasters, disaster management in all of them shared basic similarities. Disaster management was campaign-based: it relied on the existing mobilization skills of cadres and local governance structures, and it emphasized local self-help and popular participation over central assistance (*zili-gengsheng*). But in reality things were more complicated. Political patronage from the central government was crucial for successful local disaster management. Relief aid was notably frugal and people's comfort, and even their lives, were secondary to production needs. Humanitarian motives for relief aid were markedly missing. Maoist socio-political structures could be temporarily relaxed during the hazards, but upholding them was an important part of disaster governance. Campaign-based disaster management was relatively effective in the emergency phase, but it had difficulties in bringing about economic recovery after disasters.

Maoist structures of urban governance were conducive to efficient disaster management especially in emergency phases. Work units, streets and communes acted not only as specialized production units and residential organizations, but as generic mobilization units. Workers in a factory could become vegetable growers, flood fighters and reconstruction workers by party decision. The second aspect which explains effective emergency responses was the ability to use the structures of the command economy in a concentrated and coordinated manner to provide the needed supplies from outside regions. Wasteful as the command economy was in general, it could be used effectively in relatively short-term campaigns to produce and transport priority products and materials. However, this first needed authorization by the central government, which was not always forthcoming.

The willingness of the Party Centre to support disaster management was the crucial factor in many ways. It dictated how much relief aid Tianjin received from outside and whether it could tap into the national supply lines in the command economy. However, perhaps more importantly, it dictated whether the city could use its own resources for disaster management in the first place. Disaster management was always a distraction from the regular production priorities of the city. Therefore, central-level acknowledgement of disaster conditions was needed to gain permission to change priorities. In the GLF famine, this acknowledgement was missing until late summer of 1960; in 1963, the provincial and city leaders were able to make difficult decisions promptly in order to prevent the inundation of the metropolis. In 1976, disaster relief became a national priority when the new Prime Minister Hua Guofeng had to prove his capability in the eyes of other party leaders and the population who were just waiting for Chairman Mao to die.

Maoist socio-political divisions greatly affected the way vulnerabilities were generated in society. The most important artificial social division was the urban/rural divide, which translated into a highly vulnerable peasant population. In general, disaster management favoured urbanites over rural residents and peasants were treated as second-class citizens in relief aid and reconstruction, as was also the case with political outcasts with 'bad' class background. Vulnerabilities were generated also through the way the party-state sought to control private initiative both in individual charity and economic activities, which can be seen as attempts at private self-protection when people's entitlement to state-provided social protection failed to materialize.

Maoist disaster governance clearly emphasized production over living conditions or even the lives of the people. In the GLF famine, a great production campaign, the Great Leap Forward itself, caused the disaster. But even relief efforts were usually justified by enabling the people to resume production, not on humanitarian grounds, and aid was usually directed at those needed for productive purposes, not for those needing it most direly. Here one can see both the communist 'cult of production' and reverence of individual sacrifice in the struggle against nature at work. In practice it meant that those lucky enough to be in productive, as defined by the Party, since also providing leadership was 'productive', urban social positions were always favoured in disaster management.

The role of propaganda in disaster management is the final important feature of Maoist disaster management. The self-proclaimed champion of the people in all disasters was the Communist Party, even when it actually caused them, as it did in the GLF famine. Under the Party's propaganda, disasters became extremely politicized national events, dramas with happy endings. Disasters were turned into victories through propaganda. In 1963 and 1976, there was also some substance to the claims, but naturally no critical voices were heard in public. Moreover, the media offered very little practical information for residents. Indeed most disaster-related communication was either useless or outright harmful to individuals as during the GLF famine, but propaganda served the Party loyally.

All-in-all, the key to relatively effective urban disaster management in Maoist China was in its organizational preparedness. In all three disasters, the existing disaster mitigation structures, such as granaries, dykes, reservoirs, or buildings standards, were inadequate to cope with the magnitude of the hazards. However, disaster preparedness was strong in an organizational sense. And it was strong because the urban party-state was strong. A good question is how much of this preparedness has been carried over to the Reform Period.

Notes

1 DZHZH 1995, 23.
2 Chen 2012, 134.
3 Liu 2011, 8–9.
4 GCDTJ 1991, 318–19.
5 MZHJZH 1995, 264; Chen 2012, 134–6.
6 DZHZH 1995, 268–99.
7 Ibid.
8 Ibid., 275–7.
9 Ibid., 272.
10 Ibid., 270.
11 Ibid., 275. The Chinese system classifies earthquake damage on a scale of 1–12, with 12 being the most severe.
12 Ibid., 277–9.
13 Ibid., 281–2.
14 Ibid., 282.
15 Ibid.
16 Ibid.
17 Ibid., 274–5.

18 Ibid., 273.
19 The system was made famous by its apparent ability to predict a major earthquake in Liaoning in 1975. Its features are outlined by Bennett (1979). The problem with the system was that most of its predictions were false and thus useless.
20 DZHZH 1995, 271.
21 Ibid., 269.
22 Ibid., 281.
23 Contemporary reporting called it the 'Tangshan-Fengnan earthquake' based on the location of the epicentre and damage to these two regions. Officially, all together some 242,000 people died in the disaster, but this figure has been disputed (Teiwes 2007, 523).
24 Liu 2004, 342.
25 Wang Yuezhen 1996, 1.
26 Li 1996, 28.
27 Liu 2004, 355; for eyewitness accounts see e.g. Jia (2006).
28 Liu 2004, 346.
29 Qu 2006, 127.
30 DZHZH 1995, 58–9.
31 Zhong-Gong Tianjinshi weiyuanhui (29.7.1976/2004), 377.
32 DZHZH 1995, 63.
33 MZHJZH 1995, 263–4; Wang Yuezhen 1996, 1–12.
34 Chen and Booth 2011, 163.
35 TJCC 1996, 659.
36 DZHZH 1995, 284.
37 TR 22.8.1976, 2.
38 DZHZH 1995, 316.
39 Slightly different statistics are offered by MZHJZH (1995, 264). According to it, there were 24,296 dead, 21,568 seriously injured and 55,917 slightly injured. Differences in figures may be partly explained by different timeframes used. Some of the severely injured would perish from their wounds later and thus be counted differently at different times. Also clearing ruins would reveal new dead people previously not included in statistics. Why the number of injured was given so differently is difficult to tell, but the larger number may also include refugees who came to the city.
40 DZHZH 1995, 79 (Table 1.7).
41 The indirect costs are computed based on the difference of actual and budgeted city revenue each year. If the comparison is made based on 1975 income, then losses were about 1,377 billion *yuan* by the beginning of 1980 (TJCC 1966, 655–6).
42 Liu 2004, 354.
43 For eyewitness accounts see e.g. Wang Jinping 2006, 3–6; Dong 2006, 7–9; and Zhang Zhenhua 2006, 10–12.
44 Zhong-Gong Tianjinshi weiyuanhui (29.7.1976/2004), 378; Liu 2004, 354.
45 Wang Hui 2006, 168–9.
46 Liu 2011, 206.
47 Liu 2004, 346; DZHZH 1995, 312–13 and TR 4.9.1976, 1.
48 DZHZH 1995, 285.
49 Liu 2011, 206.
50 DZHZH 1995, 285–6.
51 TR 3.8.1976, 2.
52 Liu 2011, 207.
53 DZHZH (1995, 286–7) provides examples of Hangu District and Ji County; Liu Zongjian (1996, 33–40) about Heping District.
54 E.g. TR 5.8.1976, 2.
55 Wang Hui 2006, 174.
56 Wang Yuezhen 1996, 4.
57 Li 1996, 29; Wang Hui 2006, 184.

58 Wang and Shao 2006, 49–51.
59 Liu 2004, 346–7; E.g. TR 5.8.1976, 2; Hang 2006, 32–4; Liu 2011, 207.
60 TR 30.7.1976, 1.
61 TR 7.8.1976, 2.
62 TR 5.8.1976, 2.
63 TR 3.8.1976, 1; TR 4.8.1976, 1.
64 TR 4.8.1976, 2; TR 7.8.1976, 3.
65 TR 4.8.1976, 1.
66 Ibid., 2.
67 TR 11.8.1976, 3.
68 TR 9.8.1976, 3.
69 Li Zhenying, 1996, 30
70 DZHZH 1995, 308; TR 2.8.1976, 2; Wang Yuezhen 1996, 5.
71 Wang Yuzhong 1996, 41.
72 DZHZH 1995, 308; TR 2.8.1976, 2.
73 Wang Hui 2006, 175.
74 DZHZH 1995, 309 and Liu 2004, 347.
75 TR 30.7.1976, 1.
76 TR 3.9.1976, 9.
77 DZHZH 1995, 309.
78 TR 19.8.1976, 2.
79 DZHZH 1995, 313.
80 TR 21.8.1976, 2.
81 Zhong-Gong Tianjinshi weiyuanhui (1.8.1976/2004), 381.
82 DZHZH 1995, 284–6 and 313–14.
83 Ibid., 287.
84 Ibid., 288.
85 MZHJZH 1995, 264.
86 Ibid., 292.
87 Zhang Zhenhua 2006, 10–12.
88 Chen Lude 1996, 49.
89 TR 17.8.1976, 1.
90 DZHZH 1995, 288–9.
91 Ibid.; TR 3.9.1976, 9.
92 DZHZH 1995, 291.
93 Liu 2004, 350.
94 DZHZH 1995, 308.
95 TR 4.8.1976, 1.
96 Zhong-Gong Tianjinshi weiyuanhui (29.7.1976/2004), 380.
97 Ibid.
98 DZHZH 1995, 285–6.
99 Zhong-Gong Tianjinshi weiyuanhui (1.8.1976/2004), 380.
100 TR 29.7.1976, 1; the last was a reference to the campaign to criticize Deng Xiaoping that had been going on since the spring of 1976 as part of factionalist struggle in party leadership in waiting for Mao's death.
101 Zhong-Gong Tianjinshi weiyuanhui (1.8.1976/2004), 380.
102 TR 31.7.1976, 1.
103 DZHZH 1995, 302.
104 TR 30.7.1976, 4.
105 TR 5.8.1976, 2.
106 Zhong-Gong Tianjinshi weiyuanhui (6.8.1976/2004), 384.
107 Liu 2004, 343.
108 DZHZH 1995, 296.
109 TR 4.8.1976, 2.

110 TR 2.8.1976, 2.
111 Hang 2006, 32–4.
112 Zhong-Gong Tianjinshi weiyuanhui (1.8.1976/2004), 378.
113 DZHZH 1995, 286 and 294.
114 Zhang Guanghui 2006, 41–2.
115 DZHZH 1995, 286 and 294; Li 2006, 46–8.
116 Aid is described for example in TR 27.8.1976, 1; DZHZH 1995, 308.
117 Liu 2004, 348–9.
118 TR 27.8.1976, 1.
119 Liu 2004, 348.
120 Ibid.
121 TR 5.8.1976, 2.
122 See e.g. TR 3.9.1976, 9.
123 DZHZH 1995, 293–6.
124 Zhong-Gong Tianjinshi weiyuanhui (1.8.1976/2004), 382.
125 DZHZH 1995, 296.
126 Zhong-Gong Tianjinshi weiyuanhui (1.8.1976/2004), 382.
127 DZHZH 1995, 293.
128 Ibid., 296, but MZHJZH (1995, 264) gives an even higher figure of 70.4 per cent as the damage rate to residential buildings.
129 DZHZH 1995, 296.
130 TR 3.8.1976, 2.
131 Zhong-Gong Tianjinshi weiyuanhui (29.7.1976/2004), 379, exact figures of materials in Liu 2004, 348.
132 MZHJZH 1995, 264.
133 Zhong-Gong Tianjinshi weiyuanhui (6.8.1976/2004), 384.
134 On Ji County see TR 5.8.1976, 2.
135 TR 13.8.1976, 4; for a similar description of a production brigade in Wuqing County see TR 23.8.1976, 3.
136 Zhong-Gong Tianjinshi weiyuanhui (29.7.1976/2004), 379.
137 DZHZH 1995, 308.
138 Ibid., 298.
139 TR 5.8.1976, 4.
140 TR 19.8.1976, 2 and TR 6.8.1976, 2.
141 TR 6.9.1976, 3.
142 Zhong–Gong Tianjinshi weiyuanhui (6.8.1976/2004), 383–5.
143 DZHZH 1995, 295–6.
144 TR 5.8.1976, 1.
145 TR 8.9.1976, 2; TR 2.8.1976, 2; TR 6.8.1976, 2.
146 DZHZH 1995, 295–6.
147 For eyewitness accounts of life in temporary shelters see e.g. Song Xinzong 2006, 213–14; Yu Huo 2006, 226–7, and Lin 2006, 201–3.
148 Liu Xi, 1996, 45.
149 Liu 2004, 352 and DZHZH 1995, 316.
150 MZHJZH 1995, 265–6.
151 DZHZH 1995, 310.
152 The Ministry of Civil Affairs (MOCA) was re-established only in 1978. This information, taken from the Tianjin Bureau of Civil Affairs Gazetteer must refer to those units of the City Government which were in charge of refugee work that later returned to MOCA.
153 MZHJZH 1995, 264–5.
154 Ibid., 264–6.
155 Zhong-Gong Tianjinshi weiyuanhui (1.8.1976/2004), 382.
156 DZHZH 1995, 292 and 313.

157 Liu 2004, 352.
158 Ibid.
159 DZHZH 1995, 312.
160 Zhong-Gong Tianjinshi weiyuanhui (6.8.1976/2004), 384.
161 See e.g. TR 4.8.1976, 1.
162 Liu 2004, 346; TR 6.8.1976, 2.
163 TR 19.8.1976, 2.
164 DZHZH 1995, 292–3.
165 Zhong-Gong Tianjinshi weiyuanhui (1.8.1976/2004), 382.
166 Zhong-Gong Tianjinshi weiyuanhui (6.8.1976/2004), 385.
167 DZHZH 1995, 292–3.
168 Ibid., 292.
169 E.g. TR 5.8.1976, 4.
170 TR 6.8.1976, 3.
171 TR 5.8.1976, 4.
172 DZHZH 1995, 293.
173 Ibid., 285.
174 Ibid., 293.
175 Zhong-Gong Tianjinshi weiyuanhui (29.7.1976/2004), 379.
176 TR 3.8.1976, 2.
177 Zhong-gong Tianjinshi weiyuanhui (6.8.1976/2004), 386.
178 DZHZH 1995, 320; TR 20.8.1976, 4.
179 DZHZH 1995, 319.
180 TR 5.9.1976, 2.
181 Zhong-gong Tianjinshi weiyuanhui (29.7.1976/2004), 379.
182 DZHZH 1995, 320.
183 Ibid.; TR 20.8.1976, 4.
184 TR 7.8.1976, 1.
185 Ibid.
186 See e.g. Teiwes 2007.
187 TR 11.8.1976, 1.
188 Ibid.
189 Ibid.
190 TR 12.8.1976, 2.
191 TR 7.9.1976, 3.
192 TR 13.8.1976, 4; TR 15.8.1976, .
193 TR 23.8.1976, 3.
194 TR 12.8.1976, 2.
195 Ibid.
196 TR 23.8.1976, 3; a similar story also in TR 12.8.1976, 2.
197 TR 23.8.1976, 3.
198 TR 6.8.1976, 3.
199 TR 29.8.1976, 3.
200 TR 31.8.1976, 4.
201 TR 22.8.1976, 3.
202 TR 29.8.1976, 3.
203 TR 17.8.1976, 3.
204 TR 2.8.1976, 2; TR 4.8.1976, 2.
205 TR 4.8.1976, 2.
206 TR 8.8.1976, 2.
207 TR 5.8.1976, 1; TR 1.9.1976, 2.
208 TR 6.8.1976, 3.
209 TR 18.8.1976, 2.
210 TR 7.8.1976, 4.

211 TR 17.8.1976, 1.
212 TR 20.8.1976, 3.
213 E.g. TR 2.8.1976, 1.
214 TR 2.9.1976, 1.
215 TR 3.9.1976, 1.
216 DZHZH 1995, 314.
217 TR 27.8.1976, 2.
218 DZHZH 1995, 321 and MZHJZH 1995, 264.
219 Zhong-gong Tianjinshi weiyuanhui (29.7.1976/2004), 378.
220 Zhong-gong Tianjinshi weiyuanhui (31.7.1976/2004), 392.
221 TJCC 1996, 661.
222 Zhong-gong Tianjinshi weiyuanhui (29.7.1976/2004), 379.
223 E.g. TR 11.8.1976, 1.
224 TR 10.8.1976, 4.
225 E.g. TR 3.8.1976, 1.
226 Zhong-Gong Tianjinshi weiyuanhui (6.8.1976/2004), 385.
227 Zhong-Gong Tianjinshi weiyuanhui (21.8.1976/2004), 386–7.
228 Ibid., 387–9.
229 Ibid.
230 Zhong-Gong Tianjinshi weiyuanhui (25.11.1976/2004), 391.
231 Ibid., 390.
232 Ibid., 391.
233 TJCC 1996, 656.
234 Zhong-Gong Tianjinshi weiyuanhui (25.11.1976/2004), 392.
235 Zhong-Gong Tianjinshi weiyuanhui (6.8.1976/2004), 386.
236 DZHZH 1995, 315–16.
237 Zhong-Gong Tianjinshi weiyuanhui (6.8.1976/2004), 386; TR 19.8.1976, 2.
238 DZHZH 1995, 316; TR 6.8.1976, 2.
239 TR 5.8.1976, 1.
240 TR 23.8.1976, 1.
241 Ibid.
242 TR 31.8.1976, 4.
243 Wang Yuqiao 2006, 81–2; Yu Jiangyun 2006, 114–15; Yao 2006, 236–7.
244 Liu Zongjian 1996, 36.
245 Zhong-Gong Tianjinshi weiyuanhui (6.8.1976/2004), 384.
246 DZHZH 1995, 296.
247 Ibid., 297–8.
248 Zhong-Gong Tianjinshi weiyuanhui (25.11.1976/2004), 389–90.
249 Liu Xi 1996, 45.
250 DZHZH 1995, 296.
251 Ibid., 318–19.
252 TJCC 1996, 658
253 DZHZH 1995, 297.
254 Wang Hui 2006, 185.
255 TJCC 1996, 659–60; The 'small five zones' were Heping Chongrenli, Hexi Qiushili, Hongqiao Nantouyao, Xiaoxiguan and Heping Xiaotaodi.
256 Ibid., 660.
257 GCDTJ 2001, 398–401.
258 TJCC 1996, 660–2; DZHZH 1995, 325.
259 TJCC 1996, 662–3.
260 Ibid., 663.
261 Ibid., 664–5.
262 Ibid., 669–71.
263 Ibid., 663.

264 Wang Yuezhen 1996, 7.
265 TJCC 1996, 671–2.
266 Liu 2004, 353–4.
267 Zhong-Gong Tianjinshi weiyuanhui (6.8.1976/2004), 384.
268 TJCC 1996, 671–2.
269 Ibid., 671.
270 DZHZH 1995, 321.
271 Ibid., 310–11; 326–7.
272 Ibid., 322–4.
273 Liu 2004, 352.
274 Ibid., 355–6.
275 DZHZH 1995, 321.
276 Ibid., 328–9.
277 Liu 2004, 355.
278 MZHJZH 1995, 266.
279 Ibid., 266.
280 DZHZH 1995, 310–311.
281 Ibid.
282 MZHJZH 1995, 266.

Epilogue

The legacies of Maoist disaster management

After Mao died, China moved gradually into what is nowadays called the reform period. During this period, the Party has essentially abandoned Communism as the immediate goal of the regime, and scrapped most of the Maoist socio-political institutions and policies aimed at achieving this goal. Instead, the Party has increasingly embraced a market-oriented economic model. At the moment, China is in many ways a very different society than the one we have studied in this book. Gone are the mass political campaigns, official class divisions, the command economy and many structures of government such as the people's communes and revolutionary committees. The *danwei* system is seriously weakened and does not function as the basis of social organization anymore, as it did in the Maoist era. Even the *hukou* system is under reform as I write in 2015 – although urban and rural disparities will continue in the foreseeable future. China has opened its doors to the outside world and embraced the international community like never before, including accepting relief aid. Compared to before, civil society, or at least the number of civic associations, is growing in China and there is even a degree of media freedom especially on the internet. Society is much more affluent, technologically capable and urbanized than it was under Chairman Mao.

One thing, however, has not changed. Natural, biological and policy-induced hazards still occur frequently in China. If anything, natural hazards are on a slight increase in their severity (as measured by the damage they inflict) and area they cover in China. Therefore, floods, typhoons, droughts, earthquakes, epidemics and other types of hazards belong to the landscape of governance in China as tightly as ever. Bar the unlikely event of a truly global food crisis that would disturb China's food imports, the spectre of famine may have finally vanished from the Land of Famine, but as late as 1980–2 a drought in Hebei made 14 million people require food relief aid and some of the victims even resorted to eating tree bark.[1] In general, the transformation since the beginning of the reform period has changed the regime's and people's economic and technological endowments in many ways for the better in their ability to withstand natural hazards. But as discussed here, the fundamentals of the management of large natural disasters has retained many of the characteristics it had during the Maoist era. Indeed, the regime still keeps Maoist practices alive in disaster management as nowhere else in society. Arguably, this has to do with the things that have not changed, or changed only a little, during this time, namely the party-state and its style of governance.

I shall end this book with a tentative discussion on the general trends of natural disasters and disaster governance in the reform period and try to show the changes and continuities in it by briefly analysing two relatively recent major disasters/crises: the SARS epidemic in 2003 and the Wenchuan earthquake in 2008.[2] I shall then end the chapter by discussing what the Maoist experience can teach us about disaster management today.

Trends of Natural Disasters

There are some notable trends in the way natural disasters have been occurring and the damage they have caused in the reform period when compared to the Maoist and Republican periods before it, which are worth noticing. The frequency of natural disasters has remained high. In 1990–2008 natural disasters affected on average about 300 million people, destroyed 3 million buildings, forced 9 million people to evacuate their homes and caused on average 200 billion *yuan* worth of damage annually.[3] However, in historical comparison, the number of those killed by natural disasters or the ensuing secondary disasters has been going down. Omitting the GLF famine, which was arguably *sui generis* and cannot be counted as caused by natural hazards, most of the ten deadliest disasters in 1900–2009 occurred in the late Qing period, and the Republican (or warlord) period, reflecting the poor disaster management capabilities of these regimes. Only two out of ten of the deadliest disasters occurred in the Maoist period (floods in Southern China in 1959 and the Tangshan earthquake), and none of them during the reform period. However, at the same time, all of the ten costliest natural disasters took place during the reform period.[4]

These figures can be read as indicating that the contemporary Chinese state and society are much better at protecting the lives of citizens from natural disasters than the earlier regimes, but also that at the same time economic development has meant that there is simply more infrastructure and production that can suffer from disasters, thus the financial costs of damage from natural disasters have risen.

There is also some evidence that, in terms of agriculture, the area covered by natural disasters has grown during the reform period. In 1990–2010 nearly 47.36 million hectares were affected by disaster conditions (meaning drought, flood, cold, typhoons and hailstorms, in that order) and that the area has grown slightly during this period. However, areas where crops were seriously affected by these hazards were diminishing during the same period especially in terms of drought and floods. This is probably related to better disaster mitigation in terms of irrigation and other hydrological works, a large number of which have been built during the reform era around China.[5] Large sections of major (such as the Yangtze and Yellow Rivers) and middle-sized rivers have been constructed to withstand once-in-100-years-floods, and irrigation works have been improved in a number of localities. Many large-scale flood control and irrigation projects are still going on in China and Chinese authorities have also developed three-dimensional monitoring, early warning and forecasting capabilities on meteorological and hydrological events. China has launched at least five satellites specifically for the purpose and, for example, maintains nearly 25,000 meteorological observation posts.[6] Better information and disaster prevention infrastructure should also turn into better performance in disaster governance.

However, more challenges lay ahead. While it may be premature to say whether the growing acreage affected by meteorological disasters indicates the effects of global warming, some climate change scenarios estimate that drought, flooding and storms will become more severe in China in the future.[7] Even the government is officially worried about climate change and how it is likely to increase meteorological hazards.[8] Apart from more responsible climate policy by China itself, which hopefully is forthcoming sooner than later, this can only require more effective disaster governance in the future.

Another increasingly pressing current development is the continuous deterioration of China's natural environment due to pollution and mismanagement of natural resources, which has already created public health crises in many cities due to smog and other pollutants.[9] A longer discussion on the environmental problems in China is beyond the scope of this book, but it should be noted that the party-state seems to be far more efficient in dealing with sudden large natural hazards than slowly creeping public health problems, even if the latter may be far more lethal and economically damaging for the population as a whole. This of course has to do with the complexity of the pollution problem as an outcome of otherwise desirable economic activity, but also the government's own complicity with the crisis, which has been, and remains, the Achilles' heel of Chinese disaster governance. This was also highlighted in the SARS epidemic, which has been probably the single most important public health crisis during the reform period, and had its consequences for disaster governance as well.

The SARS Epidemic 2002–3

SARS, or Severe Acute Respiratory Syndrome, is a viral disease, which spread as an epidemic in late 2002 – early 2003 in China and to many other countries from there. The previously unknown type of coronavirus had a high fatality rate (almost 10 per cent of patients died) and was highly infectious through air. It is likely that the virus was transmitted to humans through zoonosis somewhere in southern China.

Although the first patients had died of SARS already in November 2002, local and central governments were slow to react to the crisis. The coronavirus itself was previously unknown, but more crucially for the epidemic was that the central government was undergoing a leadership change with Jiang Zemin stepping down from the position of Party General Secretary in favor of Hu Jintao and Premier Zhu Rongji gave way to Wen Jiabao at the 16th Party Congress and the 10th National People's Congress respectively. In the party-state system, this all took from November 2002 to April 2003 to happen. During this time, the press did not report bad news to create an impression of social stability. The central government's response was therefore to put a lid on the SARS issue and instruct hospitals in the capital not to inform the public about the true situation until the National People's Congress meeting in early April. The problem was exacerbated by local authorities' unwillingness to restrict popular movement during the Chinese New Year. Health authorities also lied publicly about the situation. At a March meeting of the State Council, a minister of public health claimed that the epidemic was under control, when it actually was spreading rapidly. When a Finnish diplomat died of SARS in Beijing in March, authorities claimed that he had been infected abroad. Only when the epidemic spread to Hong Kong in April, did the rest of the world became aware of the seriousness of the situation.[10]

The attempts to suppress information and even misinform the population and international community about the developing crisis was not of course new in light of the history of disaster management we have analysed above, but it made controlling the epidemic slow to start. Only after a Beijing military hospital doctor leaked out the real situation in the capital to the international press, did the whole picture became clearer. Before this, rumours had been circulating widely by SMS and on the internet. Under growing international pressure, the new party leadership changed its tactics in late April. Crisis management went into campaign mode with a politburo meeting on the SARS epidemic and the establishment of the National SARS Control Headquarters.[11] The new Party General Secretary Hu Jintao declared 'a war' on SARS and the health campaign was made a national priority. Authorities who tried to cover up the real situation were now threatened with severe punishments. A media campaign against SARS ensued combined with a public hygiene campaign and the establishment of special quarantine clinics for patients. May the 1st celebrations were cancelled in the capital and teaching was suspended at universities.

The measures included forcing improved information flows on the SARS situation from localities to the central health authorities and virtual mass 'detention' of the Chinese at their homes. Almost one-third of the population was required to stay at their homes or work places for between two and three weeks while being monitored for SARS symptoms. This policy was enforced not only by local street and village committees, but also work units, such as universities, which reported the SARS situation to health authorities. Although some commentators saw Chinese crisis management of SARS as being based on 'outdated' practices,[12] this mass mobilization of existing grassroots governance structures greatly helped in enforcing the anti-epidemic measures, and has been seen as key to preventing the spreading of SARS any further.[13] From the historical point of view, the crisis management approach was very similar to the Maoist period with the central level taking command of the situation and utilizing the existing grassroots level administrative organisations in carrying out a mass campaign.

In the end, SARS withered away in late spring. All in all there were about 8,000 diagnosed cases of SARS, 5,000 of them in China, with about 1,000 casualties. It is unknown whether the quarantine and public hygiene measures described above really brought the disease down, or whether this was somehow typical of an influenza cycle, but the severe health crisis did not really turn into a global pandemic and disaster this time. The incident nevertheless showed how China had already become so integrated with the rest of the world that it could not contain, or cover up, such hazards anymore. Its population was also better prepared to bypass official attempts to hide information from them, and such attempts only created distrust in the government and its ability to control the situation.[14]

It is easy to state that disaster governance was poor in the early phase of the SARS epidemic, just as it was during the GLF. The health authorities failed to inform the central government of the severity of the situation or the central government either did not realize the seriousness of the situation, or was not willing to react properly to it due to the change of leadership. However, as we have seen, Chinese disaster management also has the tradition of learning from past mistakes, and also in this case a number of measures were taken after the crisis in order to improve disaster governance. The disease control system became much better funded, rural health care reform received better attention, and information flows from localities to the central government on disaster conditions were improved.[15] The new government also announced that it was establishing a comprehensive national system to monitor and control epidemics. When the spectre of the bird flu epidemic reared its head in 2004–6, the authorities' response was already seen as much more timely and effective than during SARS.[16]

The SARS crisis also gave impetus to the improvement of disaster governance legislation. All in all, the reform period has witnessed over 30 new laws or central level regulations that deal directly or touch on disaster management,[17] but the SARS epidemic clearly triggered a new interest in the creation of comprehensive national legislation on disaster management. A series of new laws, regulations and plans were issued in the wake of the crisis including the Sudden Public Health Incident Emergency Regulations (2004); Public Emergencies Emergency Plan Framework Guide (2004); The Overall Contingency Plan of National Public Emergencies (2005); The National Natural Disaster Relief Emergency Plan (2006); and finally The Emergency Response Law (draft) in 2007. This newly established legal framework was first tested seriously in the Wenchuan earthquake in 2008.[18] Monitoring disasters for the central government also became a part of the creation of a 'public security emergency response system' launched in 2005 for handling so-called 'public security emergencies'.[19]

The way the Party connected natural disasters to other 'public security emergencies', which included for example mass protest, shows how disasters are regarded as a serious security matter, which can have dire consequences for social stability and even for the regime's prospects of staying in power. However, here one needs to note that the model for this was probably taken from the United States and its Department of Homeland Security, which since 2002 has also had the authority to manage natural disaster responses in the US.[20]

The Wenchuan Earthquake 2008

In the mid-afternoon of 12 May 2008 an earthquake measuring 7.9 on the Richter scale hit Sichuan Province. Its epicentre was located in Wenchuan, a relatively underdeveloped county in mountainous North-Eastern Sichuan with a large ethnic minority population. This was the biggest earthquake in China since the Tangshan earthquake and caused widespread devastation and casualties. It affected 417 counties in ten provinces, but the most severe damage was limited to 51 counties in Sichuan, where 21 million people lived.[21] Most of the serious damage occurred in and around Wenchuan, where the Beichuan Town was virtually levelled by the earthquake causing a quarter of total casualties. Many other townships in the region suffered similar fates.

The damage was enormous. By July 2008, 69,227 people were confirmed dead, 17,923 missing and 374,643 were injured. Moreover, 1.5 million houses were destroyed and about 6 million damaged leaving 4.8 million people homeless for a longer period of time, while an estimated 10 million people needed shelter right after the earthquake.[22] Infrastructure was also badly damaged including 34,125 km of highway, 1,263 reservoirs, 7,444 schools, 11,028 medical institutions, and numerous factories.[23] The overall cost of damage is given by different sources as being between 123.66 billion USD and 191.913 billion USD, the latter figure making it the second costliest natural disaster in history only after the Japanese 2011 Tohoku earthquake and tsunami.[24]

The earthquake revealed a number of vulnerabilities in buildings and infrastructure. Many towns were severely damaged making the disaster at least semi-urban in its nature, but also houses in rural areas suffered great damage. A major reason for such vulnerability was in many cases the disregard of earthquake-resistant building standards introduced after 1976. This proved especially fatal in the situation where towns had been constructed on an unsafe fault line and brick buildings dominated building stock in the countryside. Mountainous areas also made secondary damages, such as giant land flows, mud slides and rock falls, severe.[25]

Most notably, the earthquake toppled many schools, which were more likely to collapse than any other types of public buildings. This made schoolchildren, many of whom were unfortunately sitting in their classrooms when the earthquake hit, a notably vulnerable subgroup. 19,965 children perished in ruins and 15,000 were injured.[26] Many commercial buildings withstood the tremors much better than schools, which started rumours about 'tofu-dreg' schools where construction expenses had been saved or even diverted to other uses causing bad building quality. This provoked protests from children's parents and human rights activists including such well-known figures as Ai Weiwei, and gained international media attention.[27]

Like the SARS epidemic before it, the way the Wenchuan earthquake rescue and relief effort was governed reveals a high similarity to Maoist practices. Indeed, disaster management relied once more on campaign management practices that otherwise have become almost defunct in the reform era. As before, the central government's reaction was crucial for the rescue and relief effort and in this case it was quickly forthcoming, determined to gain results, and sustained. Because the earthquake struck in the afternoon, central government was also able to react quickly.

Right after the earthquake, Hu Jintao convened the Politburo Standing Committee, which established Earthquake Relief Headquarters in the State Council. Premier Wen Jiabao was appointed commander of the actual relief effort. Wen flew to Sichuan 90 minutes after the earthquake to oversee relief efforts in person. The Earthquake Relief Headquarters was organized into eight sub-groups: the rescue group which was composed of the representatives of the relevant departments, the army, armed police and the local party committees; the forecast monitoring group; the medical group; the life settlement group; the infrastructure group; the production recovery group; the security group; and, as always, the propaganda group.[28]

Below it in the hierarchy operated the National Disaster Reduction Committee (NDRC) of the Ministry of Civil Affairs, which acted as the coordinating body for HQ bringing together 34 relevant ministries and bureaus, including the military, for the relief effort. The Committee had been set up in 1998 under a vice-premier-level leader and has the mandate of being responsible for disaster reduction policy planning, coordinating major disaster rescue and relief activities, helping local governments in their disaster reduction work, and promoting international exchanges and cooperation in the field.[29] Other professional agencies under central government, such as the China Earthquake Administration (CEA) and the Chinese Red Cross were quick to formulate action plans and sent teams of experts to the disaster zone to help and oversee the rescue efforts.[30]

As we have seen below, such reliance on an *ad hoc* headquarters to oversee and coordinate bureaucracy to carry out disaster management work was the standard operating procedure for Maoist campaign management. In the contemporary context, professional and permanent disaster governance units exist in the central government such as the NDRC, but the organization of the Wenchuan operation shows how it was subordinated to the Party Centre and its political guidance. This time, the party leaders took the disaster seriously, which gave the Headquarters a strong mandate in its work. Another similarity to the past was the role of the military in the rescue effort. The army launched for the first time its newly drafted emergency response plan, and deployed all together 146,000 soldiers to the operation taking responsibility for the most difficult and dangerous work.[31] Rescue teams from 19 other provinces were also dispatched to Sichuan on short notice. Because of the mountainous terrain, the rescue operation had to rely on helicopters and became China's largest airlifting operation in history. Difficult weather also caused accidents and crashes during the operation.[32]

While the general features of the leadership of the emergency rescue and relief operation were very similar to the past, some other things were quite different. One of the biggest changes in post-Mao disaster governance is the role of civil society actors, NGOs and charities as well as foreign aid organizations, which have all staged a comeback in Chinese disaster management since the end of the Mao era. The earthquake triggered an unprecedented response in Chinese civil society. People gave generous donations to the victims, and NGOs, including foreign actors, rushed to the scene to take part in the rescue and relief operation, their number peaking at 263 according to one study.[33] All together 3 million volunteers were reported as having worked in the disaster area and 10 million more in other areas participating in relief work. China received 76 billion *yuan* in domestic and international donations, of which about four-fifths came from China and the rest from abroad. The government even made donations to relief aid tax-deductible.[34]

The earthquake weakened local government's governance capacity momentarily, which helped NGOs and foreign aid organizations pour in. This unprecedented civilian activism seemed to overwhelm the local authorities initially. Volunteers even rushed to take part in the difficult and dangerous rescue work. This became problematic when many such groups and individuals lacked planning, skills, tools and coordination. Some volunteer rescuers ended up being helped by professional rescuers themselves. Many local governments also displayed a marked distrust of the volunteers and their motivation, not only their skills and competence. Quickly, however, NGOs organized a number of umbrella organizations to coordinate their activities and to establish relations with the authorities. Such coordinating units helped to train volunteers for government agencies and other NGOs. They also collected donations and other resources, and supplied the rescue troops with food, medicine and water. This work has been assessed as having greatly supported the government's relief efforts.[35]

The role of foreign relief aid was also notable. Two days after the disaster, China applied for international relief aid through the UN. The response was global, with over 160 countries offering assistance to China in the form of money and resources. Japan, Russia, South Korea, Singapore, Cuba, Italy, the United Kingdom, Germany, Pakistan, France, Indonesia and the Republic of China in Taiwan, sent in rescue and/or medical teams. The Japanese rescue team was the first foreign rescue team to arrive in China after 1949.[36] The total amount of foreign assistance amounted to over 500 million USD.[37] The earthquake also gave western countries a good opportunity to voice down their criticism of China's handling of ethnic unrest in Tibet and paved the way for a more amiable show of international solidarity just before the Beijing Olympics. Such disaster diplomacy could not have taken place during the Maoist era, when the Party systematically refused foreign help for ideological reasons. Accepting foreign relief aid was one of the changes made under Deng Xiaoping in the early 1980s. There is little doubt that accepting foreign assistance makes disaster management stronger than before, and coincidently offers opportunities to improve foreign relations as well.

During the Mao era, truly civic associations did not exist. At the moment the post-totalitarian reality in contemporary China allows them, but what their role will be in disaster management is as yet open. Unlike many developing countries, China is not a weak state, where disaster management can be virtually taken over by (foreign) NGOs. The Wenchuan experience has been characterized as a state–civil society cooperation model where NGOs acted in partnership with authorities.[38] It is clear, however, that the Party will try to remain in charge of such partnerships, which would be in line with the more general reform era practice of utilizing NGOs to implement and carry out state policies.[39] It is also clear that the Party cannot simply ignore the popular enthusiasm for helping disaster victims.

Therefore it has, and will continue to try, to channel this energy for its own purposes.

The Wenchuan earthquake has been seen as an event that improved NGO standing *vis-à-vis* the state by forcing the authorities to cooperate with them and gave impetus for more institutionalized relations between the two.[40] However, while NGOs play a positive role in rescue and relief efforts and even give people outside the disaster area a channel to participate in the rescue operation and thereby give them a feeling of efficacy, NGOs can also become channels for voicing grievances created by disaster governance itself. Corruption, non-compliance with safety standards, favoritism in relief aid, and government complicity with the processes that led to disasters are all possible sources of popular discontent as they have been in the past, but now the matter of how the party-state handles NGOs has been added to the list. The internet era just makes things more difficult to control from the regime's point of view. This was demonstrated in the 'tofu-dreg' schools and the response they evoked from parents, human rights activist and even some NGOs. In Wenchuan, local governments put pressure on NGOs not to support the parents' protests.[41] The same phenomenon was visible in the way media coverage of the disaster was handled and the 'tofu-dreg schools' issue silenced after a while.

Similar to the Maoist era, the earthquake triggered a major media campaign on a national scale. The party propaganda machinery used its own outlets and instructed commercial media to produce a barrage of news coverage and commentary on the disaster and the rescue and relief effort both in print and electronic form. The rescue operation was followed by 24-hour TV coverage and the central government allowed unlimited access to the disaster area even to the foreign media, something which had not happened before. Media censorship was loosened for a while resulting in a situation where the media and public could be gathering and reporting facts on the scene more quickly and accurately than official media was able to. Even casualty data was made public, which followed a 2005 decision on the matter, but had never happened during the Maoist era. The internet was also filled with critical commentary on rescue efforts, corruption and mismanaged preparations. Indeed, some observers regarded such media coverage as unprecedented in its scale and openness.[42]

There are different explanations as to why the Party allowed such unprecedented openness in disaster coverage. It has been argued that this way it could bypass information gatekeepers on local levels and thereby exercise better control over them, which improved disaster governance. It also increased popular trust in the central government's information and helped to stamp out rumours. The essentially local nature of the disaster also made it easier to allow criticism as the shame was not on the central government. It may also have been a way to try to make people behave according to government's wishes in a disaster situation when they trusted the government better.[43] Although these reasons probably played a part in the decision to open information on the disaster to the general public, other factors also played their part. First, controlling information flows in the era of the internet, which the Maoist era could not even have dreamed of, is difficult and although the Party is fully capable of and willing to censor online information, trying to silence debates on such a major disaster might have been counterproductive to its legitimacy. Further, the 2008 Beijing Olympics were approaching quickly and the unrest in Tibet the same spring had recently strained relations with Western countries. The disaster gave a possibility to shift the focus of international attention and bring lots of sympathy to the Chinese – something which was shown in the large donations given to China from abroad.

However, while the openness of information was unprecedented, the content of the media campaign and the way it was connected to the larger policy context were quite reminiscent of the Mao era. The official propaganda in state media went viral on the meaning of the earthquake rescue and relief effort offering the 'people first policy' as the key to understanding how and why the party-state was sparing no effort in the relief and rescue campaign.[44] This was the central policy slogan of the Hu Jintao – Wen Jiabao administration at the time of the earthquake, and using it as the propaganda key in the disaster continued the practice of connecting whatever campaign was going at the time of a disaster to disaster management, which we have also seen during the Maoist era.

Many features of media coverage were indeed vintage propaganda: the media showed leaders consoling the people, picturing for example Premier Wen with tears in his eyes when talking to a child buried in ruins. The soldiers, who once again had the hardest jobs, were presented as heroes going into battle against nature. Chinese patriotism and national unity were stressed in presenting the displays of solidarity of other regions to the disaster-hit regions,[45] and the rhetoric on the role of the Communist Party caring for the people and it, and only it, leading the people to victory once again was ubiquitous. Some media quotations from the scene were also as if directly from 1963 or 1976. When 61-year-old Li Mingcui was rescued and helped out of the ruins on 19 May in Beichuan County, his first words were reported as being: 'Long live the Communist Party, thanks for the help!'[46]

Schneider and Hwang[47] have noted that the media coverage of the Wenchuan earthquake and cultural products that referred to it afterwards, such as movies, the Beijing Olympics opening ceremony, and the Shanghai Expo China's exhibition, can be seen as a form of cultural governance. One can but agree, but what needs to be added to this conclusion is that there was nothing new in this as such. Changing the changeable, the Wenchuan propaganda campaign borrowed its template and even some of its content from the Maoist era. What was different was not that propaganda/cultural governance tried to mobilize the nation in emotive and supportive responses to the regime, or the way the Party self-congratulated itself for its efforts, but that this time people were not expected to take part in the reconstruction campaign directly through a mass movement as during the Maoist era. In the reform era, less direct involvement and more indirect (i.e. financial) support is what the party-state wants.

The Wenchuan media/propaganda campaign can therefore be seen as an adaptation of the old practices to the reform-era reality, where the Party does not command all media outlets and channels of public communication as much as it did before. However, when the Party felt that it needed to control public sentiments closely, it still tried to do so. The parents of the schoolchildren who had become victims of the earthquake were paid and coerced into silence before the Olympics. Parents of the deceased children were offered about 100,000 *yuan* in compensation with about 50,000 *yuan* pensions for signing an agreement not to press their cases further. However in some localities riot police were also used to break up the protests of angry parents, and the propaganda authorities told the media to keep silent about the issue. A number of activists who continued to press the issue were given sentences for subversion. Although the government promised an inquiry into the matter of 'tofu-dreg schools', local authorities were seen as being keener to bulldoze the ruins than conduct such investigations.[48]

Also the economic context of reconstruction has changed notably from the Maoist era. The command economy has ended and the construction business is, at least in principle, based on companies operating on markets. A disaster situation is likely to impact construction markets in two ways: reconstruction creates a peak in demand for building materials and manpower causing prices to surge. Disaster may also lower local production capacity, as happened in 1976 in Tianjin, which contributes to price rises and slows down reconstruction. Both these effects were reported in the Wenchuan case, but the central government's response also had clear similarities to the Maoist era.

The central government was very keen to direct the reconstruction effort, connecting it to the general stimulus to the economy in the wake of the 2008 global financial crisis. Although much money had already been flowing to disaster areas by then, in November 2008 the government announced a 1 trillion *yuan* (about 146.5 billion USD) programme over the next three years to rebuild disaster areas. Although the reconstruction plans were revised and updated by local governments, the central government took command of the general planning. The goal was to make the rural population able to move to reconstructed houses by New Year 2009, while cities were expected to take three years to finish their reconstruction.[49]

As Yue *et al.*[50] have noted, the reconstruction plans were essentially 'post-disaster plans for reconstructing the local communities in seismically secure places and restoring the overall society to the previous state'. Therefore, reconstruction planning emphasized land use as a key to reducing vulnerabilities. Following this logic, the badly hit Beichuan town, for example, was basically relocated over 20 km away from its original location and constructed there from scratch.

The reconstruction plans were drafted in a top-down manner with officials and technical experts from various levels of government outside the disaster areas assuming the main responsibility for planning. For this, they did fieldwork at disaster sites, sometimes even enduring dangerous aftershocks, but usually only for a few weeks. This raised the concern that the planners were unfamiliar with the local situation, and that they could not possibly gather any meaningful popular feedback for the plans that sometimes expected people to move away from their old places of living to safer areas.[51] Disaster management has never had significant democratic, or more simply popular, feedback components to it in China and the Wenchuan experience continued along these lines as well.

Another aspect of reconstruction planning was that it had a clear urban bias. Although there was also a Rural Areas Reconstruction Plan, central planners devoted much more attention to city reconstruction. Rural and urban people were also given subsidies to fix and reconstruct their houses on their own when possible (urban residents received 25,000 *yuan* while rural residents were given 20,000 *yuan*), but urban reconstruction was largely driven by local governments.[52] Although the damage was naturally more extensive in the cities and the reconstruction of infrastructure was more demanding there, the bias also shows how the rural/urban divide is still alive in Chinese disaster management.

Other aspects of the reconstruction also demonstrated the lingering influences of the command economy. In principle local governments operated on construction markets acting as building contractors for reconstruction, which itself was mainly carried out by building professionals from the construction industry, no longer by people on their own and along with their real jobs. However, when housing reconstruction peaked about five months after the earthquake, it caused inflationary pressures. In response, the government tried to control the prices of major building materials in the disaster areas.[53]

The building materials needed for reconstruction included bricks, cement, steel and aggregate. For example, in February 2009 bricks cost 127 per cent more in the disaster area than before the earthquake. Although the global financial crisis of 2008 hit steel manufacturing in China hard, the industry was still reluctant to sell steel for lower prices for reconstruction and was more likely to store their products in inventories. Most building material factories in the disaster area were badly damaged and materials had to be brought in from comparatively faraway places, especially from Eastern China with its higher construction capacity. To stop the soaring prices, governments from the centre to local levels resorted to price controls with inspections of compliance. A cost-monitoring system was set up and an internet platform created for transparent comparison of prices. Sichuan authorities quickly approved the construction of 75 cement production lines, 760 brick factories and two large-scale steel mills.[54] The Ministry of Transport approved four river–land joint routes for large cargo delivery and the local authorities exempted reconstruction transportation from highway tolls.[55] Here the echoes of the Maoist disaster management practices were still visible.

The intervention into markets is said to have caused notable disincentives for producers and builders and made price speculation possible. However, reconstruction did go on with remarkable results. By early 2012, 99 per cent of post-disaster projects were reported as completed and the total cost of the relief and reconstruction effort as having cost 135 billion USD. By 2011, the Sichuan provincial economy had also recovered from the earthquake.[56]

The way reconstruction was financed was also reminiscent of the Maoist era, but had clear reform era characteristics to it, too. The central government came up with a counterpart assistance scheme whereby about 30 per cent of funds for reconstruction came from the central government and the rest from provincial and municipal governments, which were assigned their counterpart localities in worst-hit areas. Provinces and municipalities were to donate 1 per cent of their budgets to the reconstruction of their counterparts. Mobilizing sister regions to help disaster stricken areas was a Maoist practice, as we have seen, but the systematic nature of using this method in the Wenchuan reconstruction drive was something new.[57] However, charity donations, foreign aid and funding from victims themselves were other sources of funding for reconstruction and these were largely missing during the Mao-era.

China's disaster management has shown a high capability of learning from past experiences and genuinely attempting to reduce vulnerabilities in reconstruction work. In Wenchuan, the regulations laid down by the central government emphasized land use and the use of seismic-resistant materials in reconstruction. While some observers noted that only a small number of architectural changes could be detected in actually rebuilt houses,[58] the quality of reconstruction work was put under a real test in Ya'an city in Lushan County when a 6.6-magnitude earthquake hit on 20 April 2013. The earthquake killed about 200 people, but it was reported that none of the buildings built there since the Wenchuan earthquake had collapsed.[59] This would suggest that more earthquake-resistant building standards were indeed followed in the reconstruction at least in Ya'an city.

The Wenchuan earthquake also prompted the central government to formulate the new National Emergency Response Plan for Natural Disaster Relief in 2011. A major problem in 2008 was that central authorities had to make too many operational decisions which created a bottleneck in decision making, while there were also problems in coordinating civilian and military activities. The new plan outlined and divided disaster relief responsibilities among different levels of administration based on the severity of disaster, but also centralized decision making on relief aid in central government's hands. The plan was tested in practice in 2013 during the Ya'an earthquake, and has been regarded as having improved the overall leadership of disaster management from 2008.[60]

What Can We Learn from the Mao-Era Experience?

The Wenchuan earthquake generated criticism and many proposals for the improvement of Chinese disaster governance. In retrospect, it was clear that earthquake preparedness in the Wenchuan area was low. Building standards had often not been followed and public disaster awareness was not high. The local ability to respond to a disaster of this scale was limited at best. This created a typical situation in China where a centrally led rescue and relief operation, not prevention, was key to disaster management. As such there is little new in this. In all cases we have analysed in this book, some preparations against different hazards did exist, but they were always inadequate when put to the real test. Only central and local governments' high ability to quickly concentrate resources in a crisis situation compensated for this weakness. A very similar pattern is visible in Wenchuan as well.

The Wenchuan experience has given cause for demands to improve disaster governance in a number of fields, including better legislation for disaster governance and NGOs/charities, better insurance and improved pre-warning systems.[61]

Here two themes will be discussed a little more in detail: the role of NGOs and volunteers in disaster governance, and the overall disaster management system in China.

It has been argued that improved governance of civil society actors by incorporating them better into the disaster management system could strengthen it notably. It would also improve popular disaster response education and preparation and make social participation in rescue and relief better funded, more systematic and sustained.[62] However, this would require clarifying NGO's legal status in the political system, offering them channels to access the policy process, and the ability to raise funds domestically. The Party acknowledges this at least in principle. The government's *White Paper*[63] talks about improving the social mobilization mechanisms and promoting public participation in disaster prevention. It especially mentions charities and their role in providing relief aid for disaster victims.

However, it has been argued that the way the National Emergency Response Plan for Natural Disaster Relief in 2011 re-centralized disaster relief into party-state's hands left NGOs with little meaningful role in disaster-management decision making.[64] This is a potential matter of conflict, as civil society actors are more likely to assert their own agenda in disaster relief work in the future, and not only implement the party-state's policies and provide resources for its use, as the ideal situation would be from the Party's perspective. Lack of popular input in all phases of disaster management has been noted as one of the problems in Chinese disaster management,[65] and it is likely that this problem may create tensions in the future when a more affluent, better educated and informed people may want to have a larger say in disaster governance, too. Although the Party would disagree, from a Western point of view the party-state has thus far not been very successful in systematically routinizing meaningful public input in any aspect of its governance,[66] and this is therefore not a problem that concerns only disaster governance. The question is about how to manage the whole country.

Further, the idea of creating a comprehensive national disaster response system with legally defined mandates and higher levels of disaster management professionalism has also been floated by a number of scholars.[67] This is in line with the *White Paper*[68] statement that the government seeks to build a 'coordinated and efficient disaster emergency management system' in order to strengthen state capacity for emergency rescue and relief work. It has been argued that this would replace the seemingly *ad hoc* nature of first mobilizing non-expert parts of party-state administration for disaster relief and then dismantling disaster management headquarters after the event has passed every time there is a major disaster.[69] The current practice, which carries over from the Maoist era, is indeed basically temporary, and it is easy to agree in principle that such a super-bureau could improve disaster governance with professionalism, better preparedness, and possibly with better coordination among administrative agencies and civil society actors.

However, it is unclear how this would actually work in China's political system. At the moment the MOCA already has the general responsibility to plan and execute action against disasters, coordinate disaster relief and promote disaster awareness. Under it operates the National Disaster Reduction Committee, while other ministries have their hazard-specific disaster-related administrations, such as the Flood Control and Drought Relief Headquarters under the Ministry of Water Resources and the Earthquake Administration directly under the State Council. Bringing them under the same roof could make sense administratively speaking, but the peculiar nature of the party-state makes having independent professional super-agencies difficult, if not impossible. In the final analysis, the super-agency could receive its mandate only from the Party Centre, because the power to mobilize the whole administrative structure of the party-state, to disrupt the normal functioning of the economy, media, transportation and even foreign relations is so great that the Party would not relinquish it to a bureaucratic organization without keeping it under close scrutiny. Such agency could therefore never have the freedom to act on its own volition even if it had on paper a legislated mandate to do so, particularly not when the success or failure of disaster management will ultimately be attributed to the Party anyhow. As long as there is a party-state in China, the present system of the party leaders assuming *ad hoc* disaster management leadership in big natural disasters is likely to continue for good and bad.

However, another way to improve disaster management comes from the history. What the Maoist era demonstrated, was that while its physical preparedness was usually inadequate to prevent disasters or crises from occurring, organizational preparedness compensated a lot for this weakness. While the author is not proposing a return to mass mobilization based on the *danwei* and commune systems, ideological indoctrination and militias, it could be worth thinking of how grassroots civilian preparedness could be improved organizationally, especially in more unsafe regions. Creating civilian disaster preparedness teams and a national organization for them with at least basic training in disaster-related construction skills, first aid, organizing rescue and relief work, and having lines of command and rescue and disaster relief materials ready would reduce casualties and losses notably during natural hazards and prevent chaos when these occur. There has been some work to this end already,[70] but it should be a focus area of improving disaster governance, not only technology. Preferably, it should be taken in cooperation with NGOs, whose legal standing should be improved at the same time. Systematic local participation in disaster governance would further empower people in terms of their own fates in times of crisis and even improve social cohesion by providing a sense of purpose and efficacy. Maoist society is gone for good, and many people, including this author, would argue that that is basically a good thing, but something can be learned from history and the ability to learn from the past has always been a strong point in Chinese disaster management.

Notes

1 Fenby 2008, 553–4.

2 The Wenchuan earthquake is sometimes called the Sichuan earthquake after the name of the province where the epicentre of the earthquake was located, but the author uses the name Wenchuan earthquake.

3 *White Paper* 2009, 115.

4 EM-DAT: The OFDA/CRED International Disaster Database, 'Summary of Natural Disasters 1900–now' (accessed 9.7.2009).
5 Wang *et al.* 2013, 85–6; in essence the same is argued by Chen (2014, 35–6).
6 *White Paper* 2009, 123–4, 126–8.
7 http://wwf.panda.org/about_our_earth/aboutcc/problems/rising_temperatures/hotspot_map/china.cfm (accessed 1.10.2014).
8 *White Paper* 2009, 116.
9 For a brief introduction to this topic, see http://www.cfr.org/china/chinas-environmental-crisis/p12608 (accessed 1.10.2014).
10 Pan 2008; Thornton 2009.
11 Schwartz and Evans 2007, 202–4.
12 Saich 2006, 89.
13 Schwartz and Evans (2007) take generally this line, which is also argued by Kaufman (2006, 66–9).
14 Wan 2010, 191.
15 Schwartz and Evans 2007, 202–5.
16 QCD April–June 2004, 179, 853; QCD Jan–Mar 2004, 550; QCD Oct–Dec 2005, 222–3; QCD Jan–Mar 2006, 552.
17 *White Paper* 2009, 119–20.
18 Zuo 2013, 140.
19 *White Paper* 2009, 120.
20 Yao *et al.* 2011.
21 Yue *et al.* 2010, 20.
22 Wei *et al.* 2013, 95; http://www.bbc.com/news/science-environment-22398684 (accessed 1.10.2014).
23 Yan *et al.* 2010, 315.
24 Yan *et al.* 2012a, 563; http://www.bbc.com/news/science-environment-22398684 (accessed 1.10.2014).
25 Chen and Booth 2011, 19, 39 and 62–9.
26 Chen and Booth 2011, 34–5; http://www.bbc.com/news/science-environment-22398684 (accessed 1.10.2014); Teets 2009, 330.
27 Chen 2014, 38–9.
28 Zuo 2013, 140–1; Schneider and Hwang 2014, 636–7.
29 http://www.jianzai.gov.cn/ (accessed 1.10.2014); http://www.adrc.asia/nationinformation.php?NationCode=156 (accessed 1.10.2014).
30 Yan *et al.* 2012a, 563; Teets 2009, 338.
31 Zuo 2013, 143.
32 Chen and Booth 2011, 139–40.
33 Shieh and Deng 2011, 186.
34 *White Paper* 2009, 134–5; Teets 2009, 339.
35 Teets 2009; Shieh and Deng 2011.
36 Chen and Booth 2011, 201.
37 Wei *et al.* 2013, 95; Zuo 2013, 141–2.
38 Teets 2009.
39 Schwartz and Shieh 2009 on the role of the NGO in Reform Era-China in general.
40 Shieh and Deng 2011; however, Chen (2014) expresses a critical view on this.
41 Teets 2009, 345.
42 Wan 2010, 193; Zuo 2013, 141.
43 Wan 2010.
44 Schneider and Hwang 2014, 639.
45 Ibid., 645–6.
46 Zuo 2013, 143.
47 2014.

48 *New York Times* 24.7.2008, http://www.nytimes.com/2008/07/24/world/asia/24quake.html (accessed 1.10.2014); *Guardian* 18.6.2008, http://www.guardian.co.uk/world/2008/jun/18/1 (accessed 1.10.2014).
49 Yue *et al.* 2010, 18.
50 Ibid., 24.
51 Ibid., 23–4.
52 Ibid., 20; Chen and Booth 2011, 241.
53 Yan *et al.* 2010, 315; Yan *et al.* 2012b, 12–14.
54 Yan *et al.* 2010, 319–21 and 324.
55 Yan *et al.* 2012b, 13.
56 BBC 9.5.2013, http://www.bbc.com/news/science-environment-22398684 (accessed 1.10.2014).
57 Chen and Booth 2011, 229–31.
58 Yan *et al.* 2010, 322.
59 BBC 9.5.2013, http://www.bbc.com/news/science-environment-22398684 (accessed 1.10.2014).
60 Chen 2014, 40–1.
61 Yao *et al.* 2011; Zuo 2013; Shieh and Deng 2011.
62 Teets 2009, 331; Yao *et al.* 2011, 37; Shieh and Deng 2011.
63 2009, 122–3 and 133.
64 Chen 2014, 42–3.
65 Yao *et al.* 2011; Yue *et al.* 2010.
66 Fewsmith 2013 on this.
67 Yao *et al.* 2011; Zuo 2013.
68 2009, 117.
69 Yao *et al.* 2011, 36.
70 *White Paper* 2009, 122–3.

Bibliography

Secondary Sources

Ash, Robert (2006). Squeezing the Peasants: Grain Extraction, Food Consumption and Rural Living Standards in Mao's China. *The China Quarterly*, Vol. 188, 959–98.

Ashton, Basil, Kenneth Hill, Alan Piazza and Robin Zeitz (1984). Famine in China 1958–61. *Population and Development Review*, Vol. 10 No. 4, 613–45.

Bankoff, Greg (1999). A History of Poverty: The Politics of Natural Disasters in the Philippines, 1985–95. *Pacific Review*, Vol. 12 No. 3, 384–411.

Becker, Jasper (1996). *Hungry Ghosts – China's Secret Famine*. London: John Murray.

Bennett, Gordon A. (1976). *Yundong: Mass Campaigns in Chinese Communist Leadership*. Berkeley: Center for Chinese Studies, China Research Monographs, University of California.

—— (1979). Mass Campaigns and Earthquakes: Hai-Ch'eng, 1975. *The China Quarterly*, Vol. 77, 94–112.

Bernstein, Thomas P. (1984). Stalinism, Famine, and Chinese Peasants – Grain Procurements during the Great Leap Forward. *Theory and Society*, Vol. 13, 339–77.

—— (2006). Mao Zedong and the Famine of 1959–1960: A Study in Willfulness. *The China Quarterly*, Vol. 186, 421–45.

Boin, Arjen and 't Hart. Paul (2007). The Crisis Approach. In Havidán Rodríguez, Enrico L. Quarantelli and Russell R. Dynes (eds) *Handbook of Disaster Research*. New York: Springer, 42–54.

Bramall, Chris (2009). *Chinese Economic Development*. London and New York: Routledge.

—— (2011). Agency and Famine in China's Sichuan Province, 1958–1962. *The China Quarterly*, Vol. 208, 990–1008.

Bray, David (2005). *Social Space and Governance in Urban China*. Stanford: Stanford University Press, 2005.

Brown, Jeremy (2011). Great Leap City: Surviving the Famine in Tianjin. In Kimberley Ens Manning and Felix Wemheuer (eds) *Eating Bitterness – New Perspectives on China's Great Leap Forward and Famine*. Vancouver: UBCPress, 226–50.

—— (2012). *City versus Countryside in Mao's China – Negotiating the Divide*. Cambridge: Cambridge University Press 2012.

Chen, Beatrice (2005). Resist the Earthquake and Rescue Ourselves – The Reconstruction of Tangshan after the 1976 Earthquake. In Lawrence J. Vale (ed.) *Resilient City: How Modern Cities Recover from Disaster*. Oxford: Oxford University Press, 235–54.

Chen, Gang (2012). China's Management of Natural Disasters – Organizations and Norms. In Jae Ho Chung (ed.) *China's Crisis Management (3rd edn)*. London: Routledge, 130–48.

—— (2014). Institutionalization in the Politics of China's Disaster Management. *The Copenhagen Journal of Asian Studies*, Vol. 32 Nol. 1, 27–44.

Chen, Yixin (2011). Under the Same Maoist Sky: Accounting for Death Rate Discrepancies in Anhui and Jiangxi. In Kimberley Ens Manning and Felix Wemheuer (eds) *Eating Bitterness – New Perspectives on China's Great Leap Forward and Famine.* Vancouver: UBCPress, 197–225.

Chen, Yong and David C. Booth (2011). *The Wenchuan Earthquake of 2008: Anatomy of a Disaster*. Beijing: Beijing Science Press.

Cheng, Tiejun and Mark Selden (1999). The Construction of Spatial Hierarchies: China's *Hukou* and *Danwei* Systems. Teoksessa Timothy Cheek and Tony Saich (eds) *New Perspectives on State Socialism in China*. Armonk: M.E. Sharpe.

Cheng, Yong, Kam-ling Tsoi, Feibi Chen, Zhenhuan Gao, Qijia Zou and Zhanli, Chen (eds) (1988). *The Great Tangshan Earthquake of 1976: An Anatomy of Disaster*. New York: Pergamon.

Cohen, Paul A. (2003). *China Unbound: Evolving Perspectives on the Chinese Past – Critical Asian Scholarship*. London and New York: Routledge.

Cooper, Eleanor McCallie and William Liu (2002). *Grace – An American Woman in China 1934–1974*. New York: Soho Press Inc.

Dikötter, Frank (2010). *Mao's Great Famine – The History of China's Most Devastating Catastrophe, 1958–62*. London: Bloomsbury.

Dixon, John (1981). *The Chinese Welfare System 1949–1979*. New York: Preager.

Dong Zhiyong (2006). Chendiandiande hong zhihua. In Jia Chanhua (main ed.) *1976 nian da dizhen*. Tianjin: Tianjin shehuikexueyuan chuban (The 1976 Great Earthquake), 7–9.

Edgerton-Tarpley, Kathryn J. (2014). From 'Nourish the People' to 'Sacrifice for the Nation': Changing Responses to Disaster in Late Imperial and Modern China. *The Journal of Asian Studies*, 1–23 April.

Elvin, Mark (1998). Who Was Responsible for the Weather? Moral Meteorology in Late Imperial China. *Osiris*, 2nd Series, Vol. 13, 213–23.

Esherick, Joseph W (1995). Ten Theses on the Chinese Revolution. *Modern China*, Vol. 21 No. 1, 45–76.

Fan, Kawai and Lai, Honkei (2008). Mao Zedong's Fight Against Schistosomiasis. *Perspectives in Biology and Medicine*, Vol. 51 No. 2, 176–87.

Fenby, Jonathan (2008). *The Penguin History of Modern China – The Fall and Rise of a Great Power, 1850–2008*. London: Allen Lane 2008.

Fewsmith, Joseph (2013). *The Logic on Limits of Political Reform in China*. Cambridge: Cambridge University Press.

Friedman, Edward, Paul G. Pickowicz and Mark Selden (1991). *Chinese Village, Socialist State*. London: Yale University Press.

Fuller, Pierre (2011). *Struggling with Famine in Warlord China: Social Networks, Achievements, and Limitations, 1920–21*. Irvine: University of California.

—— (2013). North China Famine Revisited: Unsung Native Relief in the Warlord Era, 1920–1921. *Modern Asian Studies*, Vol. 47 No. 3, 820–50.

Gao, Hua (2011). Food Augmentation Methods and Food Substitutes during the Great Famine. In Kimberley Ens Manning and Felix Wemheuer (eds) *Eating Bitterness – New Perspectives on China's Great Leap Forward and Famine*. Vancouver: UBCPress, 171–95.

Guo Zhiying (1995) *Tianjinshi siwu nian da shiji* (Historical Records for Tianjin in the 45 Past Years). Tianjinshi Renmin Chubanshe, Tianjin.

Hoffman, Susanna and Anthony Oliver-Smith (1999). *The Angry Earth: Disaster in Anthropological Perspective*. New York: Routledge.

—— (2002). *Catastrophe and Culture: The Anthropology of Disaster*. Santa Fe: School of American Research Press.

Janku, Andrea (2012). From Natural to National Disaster: The Chinese Famine 1928–1930. In Andrea Janku, Gerrit J. Schenk and Franz Mauelshagen (eds) *Historical Disasters in Context – Science, Religion and Politics*. London, Routledge, 227–61.

—— (2013). The Internationalisation of Disaster Relief in Early Twentieth century China. *Berliner Chinahefte*, Vol. 43, 6–28.

Janku, Andrea, Gerrit J. Schenk and Franz Mauelshagen (eds) (2012). *Historical Disasters in Context – Science, Religion and Politics*. London: Routledge, 227–61.

Kaufman, Joan (2006). SARS and China's Health-Care Response: Better to be Both Red and Expert! In Arthur Kleinman and James L. Watson (eds) *SARS in China: Prelude to Pandemic?* Stanford: Stanford University Press.

Kleinman, Arthur and James L. Watson (eds) *SARS in China: Prelude to Pandemic?* Stanford: Stanford University Press.

Kung, James and Justin Yifu Lin (2003). The Causes of China's Great Leap Famine, 1959–1961. *Economic Development and Cultural Change*, Vol. 52, 51–73.

Lewis, Mark E. (2006). *The Flood Myths of Early China*. SUNY Series in Chinese Philosophy and Culture. Albany: State University of New York Press.

Li, Lillian M. (2007). *Fighting Famine in Northern China: State, Market, and Environmental Decline, 1690s to 1990s*. Stanford: Stanford University Press.

Li Ruojian (2002). Dayuejin yu kunnan shiqide shehui dongdang yu shehui kongzhi (Social Instability and Social Control during the Great Leap Forward and Period of Difficulty). *Ershiyi shiji*, Vol. 62 No. 4, 86–100.

Li, Wei and Dennis Tao Yang (2005). The Great Leap Forward: Anatomy of a Central Planning Disaster. *Journal of Political Economy*, Vol. 113 No. 4, 840–77.

Lieberthal, Kenneth G. (1980). *Revolution and Tradition in Tientsin, 1949–1952*. Stanford: Stanford University Press.

—— (1992). Introduction: the 'Fragmented Authoritarianism' Model and Its Limitations. In Kenneth G. Lieberthal and David M. Lampton (eds) *Bureaucracy, Politics and Decision Making in Post-Mao China*. Berkeley: University of California Press, 1–12.

—— (2004). *Governing China: from Revolution through Reform* (2nd edn). New York: Norton.

Lin, Yunhui (2009). *Wutuobang yundong – cong Dayuejin dao dajihuang (1958–1961)* (Utopian Campaign – From the Great Leap Forward to the Great Famine (1958–1961)). *Zhonghua renmin gongheguo shi* (The History of the People's Republic of China, Vol. 4). Hong Kong: Zhongwen Daxue Chubanshe.

Liu Shaojin (2004). *Tianjin Diqu Zhongda Ziran Zaihai Shilu* (A History of Great Natural Diasters in Tianjin Region). Tianjin: Tianjinshi Danganguan bian, Tianjin Renmin Chubanshe.

Liu Xiaorong (2011). *1966–1976 Niande Tianjin* (Tianjin in 1966–1976). Tianjin: Zhong-Gong dangshi chubanshe.

MacFarquhar, Roderick (1983). *The Origins of the Cultural Revolution 2: The Great Leap Forward 1958–1960*. New York: Columbia University Press.

—— (1997). *The Origin of the Cultural Revolution, 3: The Coming of the Cataclysm 1961–1966*. Oxford: Oxford University Press.

Meng Zhaohua (2003), *Zhongguo zainan shiji* (A History of Disasters in China) (2nd edn). Beijing: Zhongguo shehui chubanshe.

Murphy, Arthur D. (2009). Linking Broad-Scale Political Economic contexts to Fine-Scale Economic Consequences in Disaster Research. In Eric C. Jones and Arthur D.

Murphy (eds) *The Political Economy of Hazards and Disasters*. New York: Altamira Press, 3–10.

Nathan, Andrew J. (1997). Maoist Institutions and Post-Mao Period. In Andrew J. Nathan (ed.) *China's Transition*. New York: Columbia University Press, 49–62.

Naughton, Barry (2007). *The Chinese Economy –Transitions and Growth*. London and Cambridge: The MIT Press.

Oliver-Smith, Anthony (1999). What Is a Disaster? – Anthropological Perspective on a Persistent Question. In Anthony Oliver-Smith and Susanna S. Hoffman (eds) *The Angry Earth – Disaster in Anthropological Perspective*. New York: Routledge, 18–34.

—— (2002). Theorizing Disasters: Nature, Power, and Culture. In Susanna M. Hoffman and Anthony Oliver-Smith (eds) *Catastrophe and Culture: The Anthropology of Disaster*. Santa Fe: School of American Research Press.

Palmer, James (2012). *The Death of Mao – The Tangshan Earthquake and the Birth of the New China*. London: Faber & Faber.

Pan, Philip P. (2008). *Out of Mao's Shadow – The Struggle for the Soul of a New China*. New York: Simon and Schuster.

Peng, Xizḥe (1987). Demographic Consequences of the Great Leap Forward in China's Provinces, *Population and Development Review*, Vol. 13 No. 4, 639–70.

Perry, Elizabeth (2007). Studying Chinese Politics – Farewell to Revolution? *The China Journal*, Vol. 57, 1–25.

Quarantelli, E. L., Patrick Lagadec and Arjen Boin (2007). A Heuristic Approach to Future Disasters and Crises: New, Old, and In-Between Types. In Havidán Rodríguez, Enrico L. Quarantelli and Russell R. Dynes (eds) *Handbook of Disaster Research*. New York: Springer, 16–41.

Rogaski, Ruth (1997). Beyond Benevolence: A Confucian Women's Shelter in Treaty-Port China. *Journal of Women's History*, Vol. 8 No. 4, 54–90.

Ross, Lester (1984). Earthquake Policy in China. *Asian Survey*, Vol. 24 No. 7, 773–87.

—— (1983). Flood Control Policy in China: the Policy Consequences of Natural Disasters. *Journal of Public Policy* Vol. 3, Part 2, 209–32.

Rothschild, Harry N. (2012). Sovereignty, Virtue, and Disaster Management: Chief Minister Yao Chong's Proactive Handling of the Locust Plague of 715–16. *Environmental History*, Vol. 17 October, 783–812.

Saich, Tony (2006). Is SARS China's Chernobyl or Much Ado About Nothing? In Arthur Kleinman and James L. Watson (eds) *SARS in China: Prelude to Pandemic?* Stanford: Stanford University Press.

Schencking, Charles J. (2006). Catastrophe, Opportunism, Contestation: The Fractured Politics of Reconstructing Tokyo following the Great Kantô Earthquake of 1923. *Modern Asian Studies*, Vol. 40 No. 4, 833–73.

Schneider, Florian and Yih-Jye, Hwang (2014). The Sichuan Earthquake and the Heavenly Mandate: Legitimizing Chinese Rule through Disaster Discourse. *Journal of Contemporary China*, Vol. 23 No. 88, 636–56.

Schwartz, Jonathan and Gregory R. Evans (2007). Causes of Effective Policy Implementation: China's Public Health Response to SARS. *Journal of Contemporary China*, Vol. 16 No. 51, 195–213.

Schwartz, Jonathan and Shawn, Shieh (2009). *State and Society Responses to Social Welfare Needs in China: Serving the People*. London: Routledge 2009.

Sheridan, James E. (1966). *Chinese Warlord: the Career of Feng Yu-hsiang*. Stanford: Stanford University Press.

Shieh, Shawn and Guosheng, Deng (2011). An Emerging Civil Society – The Impact of the 2008 Wenchuan Earthquake on Grass-Roots Associations in China. *The China Journal*, Vol. 65, 181–94.

Shue, Vivienne (1988). *The Reach of the State: Sketches of the Chinese Body Politic.* Stanford: Stanford University Press.

—— (2006). The Quality of Mercy: Confucian Charity and the Mixed Metaphors of Modernity in Tianjin. *Modern China*, Vol. 32 No. 4, 411–52.

Strauss, Julia (2006). Morality, Coercion, and State Building in the Early PRC: Regime Consolidation and After, 1949–1956. *The China Quarterly*, Vol. 188, 870–90.

Teets, Jessica C. (2009). Post-Earthquake Relief and Reconstruction Efforts: The Emergence of Civil Society in China? *The China Quarterly*, Vol. 198, 330–47.

Teiwes, Frederick C. and Warren, Sun (1999). *China's Road to Disaster – Mao, Central Politicians, and Provincial Leaders in the Unfolding of the Great Leap Forward 1955–1959.* London: M.E. Sharpe.

—— (2007). *End of the Maoist Era: Chinese Politics During the Twilight of the Cultural Revolution, 1972–1976.* London: M.E. Sharpe.

Thaxton, Ralph A. Jr. (2008). *Catastrophe and Contention in Rural China – Mao's Great Leap Forward and the Origins of Righteous Resistance in Da Fo Village.* Cambridge: Cambridge University Press.

—— (2011). How the Great Leap Forward Ended in Rural China: Administrative Intervention versus Peasant Resistance. In Kimberley Ens Manning and Felix Wemheuer (eds) *Eating Bitterness – New Perspectives on China's Great Leap Forward and Famine.* Vancouver: UBCPress, 251–71.

Thornton, Patricia M. (2009). Crisis and Governance: SARS and the Resilience of the Chinese Body Politic. *The China Journal*, Vol. 61, 23–48.

Wan, Mingguo (2010). Government Information Management during Four Emergencies in China. *Journal of Technology Management in China*, Vol. 5 No. 3, 188–95.

Wang, Jian, Lijuan Qiao, Yueling Zhang, Junyan Zhao, Zhengjia Wang, Rehman Abdur and Xing Li (2013). The Trend Analysis on China's Agricultural Natural Risks and Improvement of the Ability of Disaster Mitigation. *Research in World Economy*, Vol. 4 No. 1, 82–94.

Wei, Jiuchang, Dingtao Zhao and Marinova Dora (2013). Disaster Relief Drivers: China and the US in Comparative Perspective. *China: An International Journal*, Vol. 11 No. 2, 93–116.

Wemheuer, Felix (2010). Dealing with Responsibility for the Great Leap Famine in the People's Republic of China. *The China Quarterly*, Vol. 201, 176–94.

Will, Pierre-Étienne and Bin R. Wong (1991). *Nourish the People – the State Civilian Granary System in China, 1650–1850.* Ann Arbor: The University of Michigan Center for Chinese Studies Publications, 1991.

Wisner, Ben, Piers Blaikie, Terry Cannon and Ian Davis (2004). *At Risk – Natural Hazards, People's Vulnerability and Disasters* (2nd edn). New York: Routledge.

Wou, Odorick Y. K. (1994). *Mobilizing the Masses – Building Revolution in Henan.* Stanford: Stanford University Press.

Yan, Chang, Suzanne Wilkinson, Regan Potangaroa and Erica Seville (2010). Interpreting Resourcing Bottlenecks of Post-Wenchuan Earthquake Reconstruction in China. *International Journal of Strategic Property Management*, Vol. 14, 314–31.

—— (2012a). Managing Resources in Disaster Recovery Projects. *Engineering, Construction and Architectural Management*, Vol. 19 No. 5, 557–80.

—— (2012b). Resourcing for Post-Disaster Reconstruction: a Comparative Study of Indonesia and China. *Disaster Prevention and Management*, Vol. 21 No. 1, 7–21.

Yang, Dali L. (1996). *Calamity and Reform in China – State, Rural Society, and Institutional Change since the Great Leap Famine*. Stanford: Stanford University Press.

Yang Jisheng (2009). *Mubei – 1958–1962 Zhongguo dajihuang jishi* (Tombstone – a Comprehensive Account of the Great Chinese Famine during the Great Leap Forward). Hong Kong: Tiandi tushu youxiangongsi.

—— (2012). *Tombstone: The Great Chinese Famine, 1958–1962*. London: Farrar, Straus and Giroux.

Yang, Nianqun (2004). Disease Prevention, Social Mobilization and Spatial Politics: The Anti Germ-Warfare Incident of 1952 and the 'Patriotic Health Campaign'. *The Chinese Historical Review*, Vol. 11 No. 2, 155–82.

Yao, Hao-wei, Zhe Zhao, Dong Liang, Zhen-zhong Chen and Arnd Rogner (2011). Preliminary Exploration on China Public Security Management System. *Procedia Engineering*, Vol. 11, 34–9.

Yu, Liu (2006). Why Did it Go So High? Political Mobilization and Agricultural Collectivization in China. *The China Quarterly*, Vol. 187, 732–42.

Yuan Xiaorong (2010). *Mao Zedong chuchu he xunshi jishi 1949–1976, xia* (Mao Zedong's Visits and Inspection Tours 1949–1976, Vol. 1). Hong Kong: Strong Wind Press.

Yue, Ge, Gu Yongtao and Deng Wugong (2010). Evaluating China's National Post-Disaster Plans: The 2008 Wenchuan Earthquake's Recovery and Reconstruction Planning. *International Journal of Disaster Risk Science*, Vol. 1 No. 2, 17–27.

Zhao, Dingxin (2009). The Mandate of Heaven and Performance Legitimation in Historical and Contemporary China. *American Behavioral Scientist*, Vol. 53 No. 3, 416–33.

Zhou, Qiuguang (2006). *Zhongguo cishan jianshi* (A Concise History of Chinese Philanthropy). Beijing: Renmin chubanshe.

Zhou, Xun (2012). *The Great Famine in China, 1958–1962 – A Documentary History*. London: Yale University Press.

Zuo, Qi (2013). The Current Situation of Sudden Natural Disaster Emergency Management of Our Government. *Canadian Social Science*, Vol. 9 No. 3, 139–45.

Gazzeteers

Caishuizhi (1995). *Tianjin tongzhi – Caishuizhi*. Tianjin: Tianjin shehuikexueyuan chubanshe.

CJConst (1987). *Tianjin tongzhi – Dangdai Tianjin chengshi jianshi*. Tianjin: Tianjin renmin chubanshe.

Dakang quzhi (1994). *Dakang Quzhi*. Tianjin shehuikexueyuan chubanshe.

DDZHGTJ Vol. 1 (1989). *Tianjin tongzhi – Dangdai Zhongguode Tianjin (xia)*. Beijing: Zhongguo Shehuikexue chubanshe.

DZHZH (1995). *Tianjin tongzhi – Dizhenzhi*. Tianjin: Tianjin Shehuikexueyuan chubanshe.

ES (2005). *Tianjin tongzhi – Tianjinshi er shang jituan*. Tianjin: Tianjin shehuikexueyuan chubanshe.

GCDTJ (1991). *Zhongguo Gongchandang Tianjin shi zuzhizhi ziliao 1920–1987*. Beijing: Zhongguo chengshi chubanshe.

GS (2001). *Tianjin tongzhi – Gong-Shang xingzheng guangli zhi*. Tianjin: Tianjin shehuikexue chubanshe.

Hangu quzhi (1995). Tianjin: Tianjin shehuikexueyuan chubanshe.

HBMZHJ (2010). *Hebei Shengzhi – di 63 juan, Minzhengzhi*. Hebeisheng difangzhi bianzuan weiyuanhui. Shijiazhuang: Hebei renmin chubanshe.

Jinghai xianzhi (1995). Tianjin: Tianjin shehuikexueyuan chubanshe.
Jinnan quzhi (1999). Tianjin: Tianjin shehuikexueyuan chubanshe.
LSJ / GB (1994). *Tianjin tongzhi – Tianjin tongzhi shangyezhi liangshijuan*. Tianjin: Tianjin shehuikexue xueyuan chuban.
MZHJZH (1995). *Tianjin tongzhi – Minzhengjuzhi*. Tianjin: Tianjin renmin chubanshe.
Shenpanzi (1999). *Tianjin tongzhi* – Shenpanzhi. Tianjin: Tianjin shehuikexueyuan chubanshe.
Tianjin Renkou (1986). *Tianjinshi renkou tongji yu diaocha*. Tianjin: Nankai daxue chubanshe.
TJCC (1996). *Tianjin tongzhi – Chengxiang jianshe zhi*. Tianjin shehuikexueyuan chubanshe.
TJDS (1995). *Tianjin tongzhi – Da shiji*. Tianjin: Tianjin shehuikexueyuan chubanshe.
TJHealth (1992). *Tianjin weisheng daji (1280–1990)*. Tianjin: Tianjin weisheng shizhi bianxiu weiyuanhui.
TJJS (1995). *Tianjin Junshi Tongzhi*. Tianjin: Tianjin shehuikexueyuan chubanshe.
TJJZ (1991). *Tianjin tongzhi – Tianjin jianzhi*. Tianjin: Tianjin Renmin chubanshe.
TJlaogai (1999). *Tianjin tongzhi – laogai-laojiaozhi*. Tianjin: Tianjin shehuikexueyuan chubanshe.
TJPrice (1997). *Tianjin tongzhi – Tianjin tongzhi wujiazhi*. Tianjin: Tianjin shehuikexueyuan chubanshe.
TJTongji (1999). *Tianjinshi tongzhi – Tianjin wushi nian 1949–1999*. Beijing: Zhongguo tongji chubanshe.
TJTZHZHQ (1996). *Tianjin tongzhi – Tianjin tongzhi zhengquan zhi zhengfu juan*. Tianjin: Tianjin Shehuikexue chubanshe.
TJZL 14 (1982). *Tianjin lishi ziliao 14 (1957–1961)*. Tianjin: Tianjin waiguoyu xueyuan yinshua chang.
Wuqing xianzhi (1991). Tianjin: Tianjin Shehuikexueyuan chubanshe.

Primary Sources

Memoirs

Chen Lude (1996). Qixin-xieli jiusi-fushang. *Tianjin wenshi ziliao xuanji* 4/1996, 48–54.
Hang Ying (2006). Dadizhen houde Tianjin cheng. In Jia Chanhua (main ed.) *1976 nian da dizhen*. Tianjin: Tianjin shehuikexueyuan chubanshe, 32–4.
Jia Chanhua (main ed.) (2006). *1976 nian da dizhen*. Tianjin: Tianjin shehuikexueyuan chubanshe.
Li Ding (2001). Shishi-qiushide mofan - huainian Wan Xiaotang tongzhi. In Han Yuxia and Tang Qi (eds) *Wan Xiaotang jinian wenji bianjizu*. Tianjin: Tianjin renmin chubanshe, 47–52.
Li Yin, Fang Ling, Zhen Min, Kai Ming, Shu Cun (1964). Guang shan wan li xin lian xin – Ji quanguo gedi dui Tianjin kanghong douzhangde juda ziyuan. In Wei Ye, Rui Yingjie, Wang Ziying (eds) *Lishixingde weida shenli – 1963 nian Tianjin Shi renmin fang xun kang hong douzheng*. Tianjin: Tianjin Renmin chubanshe, 134–46.
Li Zhengjie, Zhu Zemin, Hong Minsheng, Wang Liangjun (1964). Yi qu kaije – Ji Tianjinshi Renminde Kang Hong Douzeng. In Wei Ye, Rui Yingjie, Wang Ziying (eds) *Lishixingde weida shenli – 1963 nian Tianjin Shi renmin fang xun kang hong douzheng*. Tianjin: Tianjin Renmin chubanshe, 48–60.
Li, Zhenhua (2006). Qiangxiu diyi mianfen chang. In Jia Chanhua (main ed.). *1976 nian da dizhen*. Tianjin: Tianjin shehuikexueyuan chubanshe, 46–8.

Li Zhenying (1996). Tangu Beitang, Dongdagu dizhen jishi. *Tianjin wenshi ziliao xuanji* 4/1996, 26–32.

Li Zhongyuan (2001). Ta shishi guanxin qunzhong – shenqie huainian Wan Xiaotang tongzhi. In Han Yuxia and Tang Qi (eds) *Wan Xiaotang jinian wenji bianjizu*. Tianjin: Tianjin renmin chubanshe, 36–41.

Lin Xi (2006). Qingqing wu jia. In Jia Chanhua (main ed.) *1976 nian da dizhen*. Tianjin: Tianjin shehuikexueyuan chubanshe, 201–3.

Liu Fengcheng, Zhao Can, Guo Qilin, Yang Zhongyu, Gao Wunian, Huang Guanyu, Li Fu, and Ma Ye (1964). Wushi ge riri yeye – Tianjinshi renmin fangxun kanghong douzheng dashi jiyao. In Wei Ye, Rui Yingjie, Wang Ziying (eds) *Lishixingde weida shenli – 1963 nian Tianjin Shi renmin fang xun kang hong douzheng*. Tianjin: Tianjin Renmin chubanshe, 29–47.

Liu Xi (1996). Sheng-si yi shunjian. *Tianjin wenshi ziliao xuanji* 4/1996, 45–7.

Liu Zongjian (1996). Kangzai ziqiu rending-shengtian. *Tianjin wenshi ziliao xuanji* 4/1996, 33–40.

Qu Ming (2006). Yiming junren jinglide 30 ge xiaoshi. In Jia Chanhua (main ed.) *1976 nian da dizhen*. Tianjin: Tianjin shehuikexueyuan chubanshe, 126–8.

Song Xinzong (2006). Linjianpengli guo danian. In Jia Chanhua (main ed.) *1976 nian da dizhen*. Tianjin:Tianjin shehuikexueyuan chubanshe, 213–14.

Tao Zhenyi (2001). Xinxue he shengmin quan xiangei Dang – Wan Xiaotong tongzhide gaogui xiangsi he pinge. In Han Yuxia and Tang Qi (eds) *Wan Xiaotang jinian wenji bianjizu*. Tianjin: Tianjin renmin chubanshe, 158–84.

Tianjin Ribao reporters (1964). Yiqie weile zhansheng hongshui – ji Tianjin Shi renmin zhiyuan fangxun kanghongde douzheng. In Wei Ye, Rui Yingjie, Wang Ziying (eds) *Lishixingde weida shenli – 1963 nian Tianjin Shi renmin fang xun kang hong douzheng*. Tianjin: Tianjin Renmin chubanshe, 110–21.

Wang Gangzhi, Zhang Zhisheng, Xiao Lianpei (1964). Tong qiang tie bi. In Wei Ye, Rui Yingjie, Wang Ziying (eds.). *Lishixingde weida shenli – 1963 nian Tianjin Shi renmin fang xun kang hong douzheng*. Tianjin: Tianjin Renmin chubanshe, 61–73.

Wang, Hui (2001). "Wo yi wo xue jian jinmen" – mianhuai Wan Xiaotang tongzhi. In Han Yuxia and Tang Qi (eds) *Wan Xiaotang jinian wenji bianjizu*. Tianjin: Tianjin renmin chubanshe, 120–32.

—— (2006). Tianjin dizhen zaihai wenhua jixiang. *Tianjin wenshi ziliao xuanji* 2/2006, 168–78.

—— (20.7.2009). *Gensui shiwei shuji xiaxiang kaocha ji (zhuan)*. http://blog.sina.com.cn/s/blog_4ac30c600100er8y.html (accessed 4.6.2014).

Wang Jinping (2006). Weinan zhi zhongde zhenqing. In Jia Chanhua (main ed.) *1976 nian da dizhen*. Tianjin: Tianjin shehuikexueyuan chuban, 3–6.

Wang Yuezhen (1996). Kangzai jiuzai chongjian jiayuan. *Tianjin wenshi ziliao xuanji* 4/1996, 1–10.

Wang Yuqiao (2006). Zhenliu: Feichang shiqi mingci. In Jia Chanhua (main ed.) *1976 nian da dizhen*. Tianjin:Tianjin shehuikexueyuan chubanshe, 81–2.

Wang Yuzhong (1996). Nanwangde 1976 nian. *Tianjin wenshi ziliao xuanji* 4/1996, 40–4.

Wei Ye, Rui Yingjie, Wang Ziying (1964). Lishixingde weida shenli – 1963 nian Tianjin Shi renmin fang xun kang hong douzheng. In Wei Ye, Rui Yingjie, Wang Ziying (eds) *Lishixingde weida shenli – 1963 nian Tianjin Shi renmin fang xun kang hong douzheng*. Tianjin: Tianjin Renmin chubanshe, 1–10.

Wang Xiuying and Shao Yunfu (2006). Kejingde jiedao ganbu. In Jia Chanhua (main ed.) *1976 nian da dizhen*. Tianjin: Tianjin shehuikexueyuan chubanshe, 49–51.

Xie Yan (2011). *Wan Xiaotang dailing Tianjinren jie liang-duhuang*. http://www.xywq.com/grwj_article_show.aspx?id=2323, 2011.12.17 (accessed 23.4.2014).
Xing Guojun and Jie Zhengli (2006). Na nian, wo zai Ninghe dang shuji. In Jia Chanhua (main ed.) *1976 nian da dizhen*. Tianjin:Tianjin shehuikexueyuan chubanshe, 64–6.
Yao Weitong (2006). Gege bei "zhenliu". In Jia Chanhua (main ed.) *1976 nian da dizhen*. Tianjin: Tianjin shehuikexueyuan chubanshe, 236–7.
Yu Huo (2006). Linjian pengli qu qi sheng zi. In Jia Chanhua (main ed.) *1976 nian da dizhen*. Tianjin: Tianjin shehuikexueyuan chubanshe, 226–7.
Yu Jiangyun (2006). Tianjin zhiqing mianpiao hui jin. In Jia Chanhua (main ed.) *1976 nian da dizhen*. Tianjin:Tianjin shehuikexueyuan chubanshe, 114–15.
Zhang Guanghui (2006). Quanli wei fu lao dabing. In Jia Chanhua (main ed.) *1976 nian da dizhen*. Tianjin: Tianjin shehuikexueyuan chubanshe, 41–2.
Zhang Zhenhua (2006). Qiangqiu nü laoshi. In Jia Chanhua (main ed.) *1976 nian da dizhen*. Tianjin: Tianjin shehuikexueyuan chubanshe, 10–12.

Abbreviations of Collections/Journals

FXZL (1963). *1963 Nian Tianjinshi fangxun gongzuo ziliao*. Tianjin: Tianjin Shi Fangxun zhihuibu 1963.

TCA Tianjin City Archive

TJZHB *Tianjin shizheng zhoubao* (before 1961), *Tianjin Zhengbao* (after 1961), the Tianjin City Bulletin

TR *Tianjin ribao*

QCD *Quarterly Chronicle and Documentation*. The China Quarterly, 2004–6

Documents in Collections

CCP Central Committee (27.6.1981/1991). A Resolution on Certain Questions in the History of Our Party since the Founding of the People's Republic of China, 27.6.1981. In *Major Documents of the People's Republic of China (December 1978–1989)*. Beijing Foreign Languages Press.
Jiuzai jianbao No. 15 (9.8.1963). Guanyu anzhi fenhongqu zaimin zhujian tiaozheng qingkuang, 39.
Mao Zedong (Tse-tung) (1945a) On Coalition Government. In *The Selected Works of Mao Tse-tung III*, April 24.
Mao Zedong (Tse-tung) (1945b) Policy for Work in the Liberated Areas for 1946. In *The Selected Works of Mao Tse-tung IV*, December 15.
Mao Zedong (Tse-tung) (1950) Fight for Fundamental Turn for the Better in the Nation's Financial and Economic Situation. In *The Selected Works of Mao Tse-tung V*, June 6.
Mao Zedong (Tse-tung) (1955a) On the Co-Operative Transformation of Agriculture. In *The Selected Works of Mao Tse-tung V*, July 31.
Mao Zedong (Tse-tung) (1955b) Editor's Notes from Socialist Upsurge in China's Countryside. In *The Selected Works of Mao Tse-tung V*, September and December.
Tianjinshi fangxun zhihuibu (1963/1999a). Tianjin Shi fangxun zhihuibu guanyu jiajiang Tianjin Shi fanxun gongzuo gei Hebei Sheng fangxun zhihuibude baogao, *Tianjin Shi fangxun kanghong douzheng shiliao xuan. Tianjin dang'an shiliao*, No. 1, 43–4.

Tianjin Shi fangxun zhihuibu (1963/1999b). Tianjin Shi fangxun zhihuibu guanyu dangqian fanxun douzheng qingguan he jinhou bushu gei Hebei Sheng fangxun zhihuibude baogao. *Tianjin dang'an shiliao*, No. 1, 45–8.

Tianjin Shi fangxun zhihuibu (1963/1999c). Tianjin Shi fangxun zhihuibu guanyu dangqian fanxun douzheng qingguan he jinhou bushu gei Hebei Sheng fangxun zhihuibude baogao. *Tianjin dang'an shiliao*, No. 1, 48–52.

Wan Xiaotang (9.11.1960/2001). Geji dangwei mishu yao renzhen zhuahao qunzhong shenghuo. In *Wan Xiaotang jinian wenji bianjizu*. Tianjin: Tianjin renmin chubanshe, 274–6.

White Papers of the Chinese Government (2009–2011). Information Office of the State Council of the People's Republic of China. Beijing: Foreign Language Press, 2012.

White Paper (2012). China's Actions for Disaster Prevention and Reduction (May 2009). In *White Papers of the Chinese Government* (2009–2011), Information Office of the State Council of the People's Republic of China. Beijing: Foreign Language Press, 113–40.

Zhong-Gong zhongyang (4.7.1959). Zhong-Gong zhongyang guanyu zai da-zhong chengshi fazhan fushipin shengchande zhishi (1). In *Jianguo yilai zhongyao wenxian xuanbian* Vol. 11. The Database of the CPC http://58.68.145.22/directLogin.do?target=cpc (accessed 3 October 2013).

Zhong-Gong Tianjin Shi weiyuanhui (29.7.1976/2004). Zhong-Gong Tianjin Shi weiyuanhui guanyu wo shi dizhen he kangzhen jiuzai qingkuangde huibao tigang. In Liu Shaojin (ed.) *Tianjin diqu zhongda ziran zaihai shilu*. Tianjin: Tianjin Renmin Chubanshe, 377–9.

Zhong-Gong Tianjin Shi weiyuanhui (1.8.1976/2004). Zhong-Gong Tianjin Shi weiyuanhui guanyu kangzhen jiuzai qingkuangde di er ci baogao. In Liu Shaojin (ed.) *Tianjin diqu zhongda ziran zaihai shilu*. Tianjin: Tianjin Renmin Chubanshe, 380–2.

Zhong-Gong Tianjin Shi weiyuanhui (6.8.1976/2004). Zhong-Gong Tianjin Shi weiyuanhui guanyu kangzhen jiuzai qingkuangde di san ci baogao. In Liu Shaojin (ed.) *Tianjin diqu zhongda ziran zaihai shilu*. Tianjin: Tianjin Renmin Chubanshe, 383–6.

Zhong-Gong Tianjinshiwei yanjiushi (31.7.1976/2004). Zhong-Gong Tianjinshiwei zhengce yanjiushi: 1976 nian quanshi gongchang qiye zhensun qingkuang jianxun. In Liu Shaojin (ed.) *Tianjin diqu zhongda ziran zaihai shilu*. Tianjin: Tianjin Renmin Chubanshe, 392–7.

Zhong-Gong Tianjin Shi weiyuanhui (21.8.1976/2004). Zhong-Gong Tianjin Shi weiyuanhui guanyu kangzhen jiuzai qingkuangde di si ci baogao. In Liu Shaojin (ed.) *Tianjin diqu zhongda ziran zaihai shilu*. Tianjin: Tianjin Renmin Chubanshe, 386–9.

Zhong-Gong Tianjin Shi weiyuanhui (25.11.1976/2004). Zhong-Gong Tianjin Shi weiyuanhui guanyu shiyiyue shiwuri zhenzai qingkuangde baogao. In Liu Shaojin (ed.) *Tianjin diqu zhongda ziran zaihai shilu*. Tianjin: Tianjin Renmin Chubanshe, 389–92.

FXZL

FXZL (25.5.1963). Zhongyang fangxun zhihuibu guanyu jishi zuohao fangxun zhunbei gongzuode jinji tongzhi. 1–2.

FXZL (30.5.1963). Hebei sheng 1963 nian fangxun gongzuo jihua. 7–11.

FXZL (21.6.1963). Zong-Gong Hebeisheng weiyuanhui, Hebeisheng renmin weiyuanhui: pizhuan sheng fangxun zhihuibu guanyu fangxun huiyi qingkuangde baogao. 3–6.

FXZL (24.7.1963). Pizhuan shi fangxun zhihuibu guanyu jinyibu zuohao fangxun gongzuode baogao. 26–7.

FXZL (8.8.1963). Tianjin Shi Zhong-Gong weiyuanhui, Tianjin Shi renmin weiyuanhui: Zhaokai fangxun gongzuo huiyi shang Wang Peiren fushizhangde baogao. 29–31.
FXZL (10.8.1963). Hebeisheng Zhong-gong weiyuanhui, Hebeisheng renmin weiyuanhui: guanyu fangxun jiuzai gongzuode jinji zhishi (zhaiyao). 18.
FXZL (13.8.1963). Guanyu fangxun kanghong douzheng jinji zhishi. 35–6.
FXZL (16.8.1963). Tianjin Shi fangxun zhihuibu, guanyu zuzhi fangxun duiwu jige wentide tongzhi. 207–8.
FXZL (23.8.1963). Tianjin Shi fangxun zhihuibu, Tianjin Shi laodongju: guanyu fangxun kanghong qijian youguan gongzi zhifu wentide lianhe tongzhi. 212.
FXZL (1.9.1963). Tianjin fangxun zhihuibu guanyu zuohao fangxun duiwu diaozheng gongzuode tongzhi. 209–10.
FXZL (15.9.1963). Hebei Sheng fangxun zhihuibu: Guanyu dangqian fangxun gongzuode ji dian bushu. 19–20.
FXZL (22.9.1963). Hebei renmin weiyuanhui: guanyu dui zai kanghong qianjianzhong shangwang renyuan wubei fuxu wentide tongzhi. 226–7.
FXZL (25.9.1963a). Duliujianhe fangxun kanghong douzheng zongjie. 101–9.
FXZL (25.9.1963b). Machangjianhe fangxun zhihuibu gongzuo zongjie. 133–7.
FXZL (28.9.1963). Shenghuo gongying gongzuo zongjie. 264–9.
FXZL (30.9.1963a). Ziyahe fangxun kanghong douzheng zongjie. 110–20.
FXZL (30.9.1963b). Shanggulin fangxun zhihuibu gongzuo zongjie. 138–9.
FXZL (30.9.1963c). 1963 Nian yuqing shuiqing. 85–100.
FXZL (3.10.1963). Tianjinshi fangxun zhihuibu: guanyu zai fangxun zhengdou zhong huisun qingmiao, jianzhu chulide tongzhi. 349.
FXZL (15.10.1963a). Dongyuan zuzhi gongzuo zongjie. 190–3.
FXZL (15.10.1963b). Liangwanzhuang fangxun zhihuibu gongzuo zongjie. 122–3.
FXZL (8.12.1963). Jinji yundong yuan qilai, xiang hongshui douzheng – Zong-Gong Tianjin Shi weiyuanhui gei shidangyuande yi feng xin. 42–3.
FXZL (1963). Hebeisheng kanghong lieshi mingdan. 369–70.

TJZHB

TJZHB 21 (26.5.1958). Liangshibumen dayuejin hou xiang yi xin. 6–7.
TJZHB 24 (16.6.1958). Weihu shiliang tonggou-xiao zhengce, jiaqiang kongzhi shiliang wailiu. 13–14.
TJZHB 45 (10.11.1958). Ge danweide huoshituan bu yao zhijie xiang nongcun renmin gongshe lai gou shucai. 9.
TJZHB 1 (5.1.1959). Shi renmin weiyuanhui guanyu diaozheng mianfen dapei gongying bizhongde tongzhi. 10.
TJZHB 7 (16.2.1959). Shi renmin weiyuanhui guanyu yanjin ge danwei zhijie xiang nongcun renmin gongshe lai gou shucai, zhurou, deng fushipinde tongzhi. 9–10.
TJZHB 11 (16.3.1959). Shi renwei fachu zhishi, fadong qunzhong zhongzhi bima, xianrikui zengchan youke. 8–9.
TJZHB 14 (6.4.1959). Yange zhi zhizhi nongcun laodongli mangmu wailiu. 9–10.
TJZHB 18 (4.5.1959). Zhong-Gong Tianjin shiwei, Shi renmin weiyuanhui guanyu jiaqiang dangqian shichang wujia guanlide ji xiang guiding. 2–5.
TJZHB 22 (1.6.1959). Zhong-Gong Tianjin shiwei, Shi renmin weiyuanhui guanyu jiza dongshou zhunbei qieshi zuohao xiashou gongzuode jinji zhishi. 2–4.

TJZHB 23 (8.6.1959). Yao dongyuan mangliu rushi neide nongcun cheliang huanxiang shengchan. 10.
TJZHB 24 (15.6.1959). Zhong-Gong Tianjin shiwei, Shi renmin weiyuanhui qieshihao xialiang zhenggou he tongxiao gongzuode zhishi. 2–3.
TJZHB 26 (29.6.1959). Shi renmin weiyuanhui guanyu jiaqiang fu shipin guanlide ji xiang guiding. 3–4.
TJZHB 28 (13.7.1959). Shi renmin weiyuanhui guanyu yasuo shi gongyong bu, jieyue gongye yong bu he shidang kuangda renmin yong bu shoupiao fanweide zhidao. 3.
TJZHB 30 (27.7.1959). Guzuganjin, xianqi cengchan-jieyue xin gaochao yingjie guoqing shi zhounian. 6–12.
TJZHB 33 (17.8.1959). Zhong-Gong Tianjin shiwei, Shi renmin weiyuanhui guanyu shanshi-shanzhongdi zuohao xialiang zhenggou gongzuode zhishi (zhaiyao). 5–7.
TJZHB 36 (7.9.1959). Zhong-Gong Tianjin Shi weiyuanhui fachu tongzhi yange kongzhi zengjia xin renyuan. 7.
TJZHB 39 (28.9.1959). Huanhu Tianjin shi nian laide weida jianshe chengjiu. 4–12.
TJZHB 45 (9.11.1959a). Jianjue shixian woshi fushipin quanmian zigeide fangzhen. 8–9.
TJZHB 45 (9.11.1959b). Jieshao yi ge yangzhu duo, zhiba duo, liangshi cengchan duo, sheyuan shouru duode shengchandui. 14–15.
TJZHB 52 (28.12.1959). Yao zhuyi shucaide jieyue he shichang guanli. 5–7.
TJZHB 2 (11.1.1960). Shi renmin weiyuanhui guanyu chengli shi fushipin guanli weiyuanhuide tongzhi. 4.
TJZHB 4 (25.1.1960). Guanyu Chunjie shichang shichang gongyingde anpai. 3.
TJZHB 15 (11.4.1960a). Shi renminweiyuanhui guanyu jixu kaizhan shehui jieyue yundongde baogao. 8–9.
TJZHB 15 (11.4.1960b). Cong ge fangmian jieyue liangshide xiaofei. 11–12.
TJZHB 15 (11.4.1960c). Hexiqu Huiduijie zhengdun shiliang dingliang hou bian queliang wei yuliang. 12–13.
TJZHB 15 (11.4.1960d). Tiangang yi chang ban hao shitang zhichi shengchan. 13–14.
TJZHB 15 (11.4.1960e). Hexiqu Huiduijie zhengdun shiliang dingliang hou bian que liang wei xiang. 12–13.
TJZHB 15 (11.4.1960f). Qiandezhuang tonglu jumin chengli yongliang huzu jihua yong liang, jieyue liangshi chengji xianzhu. 14–15.
TJZHB 17 (25.4.1960). Liangdian yu shitang heyi liangshi gongying gongzuode xin fazhan. 6–7.
TJZHB 18 (2.5.1960). Chengsheng-qianjin, dagan wu-liuyue, wei shixian nongye da fengshou er fendou. 11–12.
TJZHB 19 (9.5.1960). Ge xian da zhong guacaide qunzhong yundong jinbu kuai, zhiliang hao. 6–7.
TJZHB 26 (27.6.1960): Zhuazhu mai shou jijie, da gao maixie zonghe liyong. 8–9.
TJZHB 27 (4.7.1960a). Dagao jiguan, qiye, fushipin shengchan, wei xunsu shixian fushipin zigei er fendou. 2–5.
TJZHB 27 (4.7.1960b). Wei nongye shengchan fuwu, wei kanghan fuwu, wo shi duozhong jingji shengchan pengbo fazhan. 9–10.
TJZHB 27 (4.7.1960c). Dagao liangshi fushengchanpin zonghe liyong, wei guojia chuangzao geng duode caifu er nuli. 8–10.
TJZHB 32 (8.8.1960a). Guanyu jin yibu kaizhande jieyue liangshi yundong de yijian. 4–5.
TJZHB 32 (8.8.1960b): Quan dang quan min dongyuan, tuji qiangzhong qiucai. 6–8.

TJZHB 32 (8.8.1960c). Shirenweiyanhui bangongting guanyu shiji jiguan fushipin shengchan youguan wentide tongzhi. 6.
TJZHB 35 (29.9.1960). Ganbu jie liang shangjiao banfa. 10.
TJZHB 37 (12.9.1960). Guanyu jixu shenru guanche yi nongye wei jichude fangzhen quan renmin daban nongye, daban liangshide yijian. 2–5.
TJZHB 40 (3.10.1960a) Zhong-Gong Tianjin wei, shi renmin weiyuanhui lianhe fachu jieyue shiyong qiyou, chaiyoude jinji tongzhi. 2–3.
TJZHB 40 (3.10.1960b). Shi renmin weiyuanhui guanyu jinzhi cong nongcun, waidi diaoru renyuande tongzhi. 2.
TJZHB 30 (25.7.1960). Zhujin shiji, da zhong qiucai. 8–9.
TJZHB 43 (7.11.1960). Guanyu jintian qiucaide fenpei he chuzang wenti. 12.
TJZHB 44 (21.11.1960a). Shi renmin weiyuanhui guanyu Tianjin jiujing chang, Tianjin juanyin chang weifan guojia zhengce, songzi dao nongcun taijia qianggou fushipinde tongbao. 2.
TJZHB 44 (21.11.1960b). Shi renmin weiyuanhui fachu guanyu chongshen qianfa ”guanyujiaqiang dangqian fushipin shichang wujia guanlide zhishi” de tongzhi. 3–4.
TJZHB 45 (5.12.1960a). Guanyu jin dong ming chun kaizhan renmin weisheng baojian yundongde yijian. 6–7.
TJZHB 45 (5.12.1960b). Shi renmin weiyuanhui fachu guanyu jiaqiang shiqu liangshi guanli jigoude guiding. 10.
TJZHB 45 (5.12.1960c). Guanyu xuejiao xitong zhong dagao qunzhong shengminde qingguan baogao. 6–7.
TJZHB 45 (5.12.1960d). Shi renmin weiyuanhui guanyu jianjue, zhizhi qingke, songli, song jinqi, he tingzhi xinjiang zhaodaisuode tongzhi. 9–10.
TJZHB 1 (17.1.1961a). Shi renmin weiyuanhui pizhuan shi Gonganju, Liangshiju ‘Guanyu baowei quandang daban nongye, daban liangshi, jin yi bu jiaqiang hukou guanlide yijian’ de tongzhi. 2–5.
TJZHB 1 (17.1.1961b). Shi renminweiyuanhui guanyu jiguan, tuanti, gong-kuang qiye, budui, xuexiao deng danwei ziyang shengzhu keyi zi zai zi shide tongzhi. 6.
TJZHB 5 (22.3.1961). Guanyu kaizhan yi fangzhi jibing wei zhongxinde chunjie weisheng yundongde yijian. 5–7.
TJZHB 6 (31.3.1961). Renzhen zhixing jin jiao nongye vshengchan yi cai wei gangde fangzhi, xianqi chuncai shengchan gaochao. 2–5.
TJZHB 11–12 (30.6.1961a). Tianjin Shi jiguan, tuanti deng kaiken jinjiao huangdi shixing banfa. 2–3.
TJZHB 11–12 (30.6.1961b). Shi renwei fachu tongzhi zhizhi shimin he zhigong suiyi dao nongcun goajia shougou shucai. 8–9.
TJZHB 19 (20.10.1961). Zhong-Gong Tianjin shiwei, shi renmin weiyuanhui guanyu jiji kaiken jinjiao huangdide zhishi. 2–4.
TJZHB 20 (31.10.1961a). Zhong-Gong Tianjin Shi weiyuanhui Tianjinshi renmin weiyuanhui guanyu zuohao qiucai, xunzai, yunshu he fenpei gongzuode tongzhi. 3.
TJZHB 20 (31.10.1961b). Wushi zonghe shangdian xuxin tingqu qunzhong yijian renzhen jingai shucai gongying gongzuo. 6.
TJZHB 21–22 (30.11.1961). Shi renmin weiyuanhui guanyu shixing ”Jin jiaobu tonggou shiliang, youliao, shucai, jiang shou gongyepin banfa” de tongzhi. 2–3.
TJZHB 24 (25.12.1961). Zhong-Gong Tianjin wei, shi renmin weiyuanhui guanyu jinyibu zhengcun yu fazhan jiguan nongfuye shengchande zhishi. 2–4.
TJZHB 1 (12.1.1962). Jiaqiang shiliang guanli, baozheng heli gongying. 7.

TJZHB 3 (10.2.1962). Tianjin Shi Cai-zheng-mao weiyuanhui guanyu dangqian shichang gongying he renmin shenghuo anpai qingkuangde baogao. 8.
TJZHB 8–9 (12.5.1962). Tianjin Shi jinjiaoqu 1962 nian cainong kouliang buzhu he chao'e jiaoshou shanpincaide jiangli banfa. 6–8.
TJZHB 16 (31.8.1962a). Shi renmin weiyuanhui guanyu jinjiaobu ji zhong nongfuchanpin shougou he jiangshoude ji xiang guiding. 2–3.
TJZHB 16 (31.8.1962b). Renzhen zuohao jiancha hedui woshi renkou, gongzhong he liangshi dingliangde gongzuo. 4.
TJZHB 16 (31.8.1962c). Zhizhi jin jiaobu shengchandui ziyou chushou shucai he cheng-xiang shangfan dao jin jiao fanyun shucai. 9.
TJZHB 18 (15.10.1962). Shi renmin weiyaunhui guanyu zan ting fafang "fushipin gouhuo zheng" de tongzhi. 4–5.
TJZHB 22 (26.11.1962). Shi renmin weiyuanhui guanyu yanke kongzhi xin zhāoshōu renyuande tongzhi. 4.
TJZHB 11 (29.6.1963). Tianjin Shi 1963 nian fangxun gongzuo jihua. 3–5.
TJZHB 16 (31.8.1963a). Zhong-Gong Tianjin shiwei, shi renmin weiyuanhui jinji zhishi quanli gonggu tifang, zhengqu kanghong douzhengde juedingxing shenli. 2–3.
TJZHB 16 (31.8.1963b). Shi renmin weiyuanhui guanyu anzhi fen hong qu nongmin he shourong jiuji zaiminde gongzuo jihua (zhaiyao). 4–6.
TJZHB 16 (31.8.1963c). Zhong-Gong Tianjin shiwei, shi renmin weiyuanhui jinji zhishi quanli gonggu tifang, zhengqu kanghong douzhengde juedingxing shenli. 5.
TJZHB 19 (1.11.1963a). Shi renminweiyuanhui bangongting guanyu jieyue kouliang zhiyuan zaiqude tongzhi. 10.
TJZHB 19 (1.11.1963b). Zhizhi ge qiye suiyi zhaoyong zaiqu nongmin. 7.
TJZHB 21–22 (16.12.1963). Shi renminweiyuanhui guanyu jin yi bu zuohao fenhongqu zaimin anzhi he shengchan juzai gongzuode zhishi. 3–7.
TJZHB 23–24 (31.12.1963). Hebei Sheng Tianjin Shi renminweiyuanhui gongzuo baogao (zhaiyao). 2–17.

TCA

TCA X0095-C-000754-048 (May 1959). Tianjin Shi Liangshiju: Tianjin Shi Hepingqu liangshi lingshou gongsi xianjin shiji.
TCA X0095-C-000754-04 (22.5.1959). Tianjin Shi Liangshiju: Hongqiaoqu liangshi lingshou gongsi xianjin shiji.
TCA X0283-c-000868-069 (1.6.1959). Tianjin Shi renmin shenghuo duo zhong jingji shengchan weiyuanhui: guanyu nongcun bufen qunzhong fasheng fuzhong ji zhiliao qingkuang xiang shiweide baogao.
TCA X0096-C-000026-012 (15.6.1959). Song Jingyi fushizhang zai Tianjin Shi zhaokaide qiucai chan-xiao huiyi shang suo zuode dahui baogao.
TCA X0096-C-000026-012 (1.7.1959). Tianjin Fushipinju: Tianjin Shi 1959 nian baicai fangbing zengchan jishu cuoshi (cao'an).
TCA X0095-C-000754-048 (1.9.1959). Tianjin Shi Liangshiju: guanyu zengchan-jieyue hongqi jingsai yundong kaizhan qingkuangde baogao.
TCA X0196-C-001086-037 (10.10.1959). Tianjin Shi Liangshiju: guanyu di san jidu zengchan-jieyue hongqi jingsai yuandongde jianjie he di si jidu gongzuo anpai yijian.
TCA X0095-C-000754-048 (25.12.1959). Zhong-Gong Tianjin shiwei, Tianjin shi renmin weiyuanhui guanyu zhaokai yumi ganshu shengchan yuejin huiyide tongzhi.

TCA X0096-C-000026-012 (15.11.1959). Tianjin Shi Fushipinju: dagan yi dong yi chun, wei hen kuaidi shixian "baozheng shuliang, duo zhong duo yang, yue junheng gongying, sijie changing" er douzheng!
TCA X0003-Y-008983 (16.1.1960). Zhong-Gong Tianjin shiwei: guanyu muqian nongcun renkou mengliude qingkuang baogao.
TCA X0053-C-001746-003 (7.3.1960). Tianjin Shi gonggong weishengju: guanyu noncun dangqian fuzhongbing qingkuang baogao 60 hao.
TCA X0003-Y-009036 (22.4.1960). Zhong-Gong Shanxi Zhunci shiwei.
TCA X0003-Y-009036 (27.4.1960). Zhong-Gong Wenshui Xian shi yuanhui.
TCA X0003-Y-009036 (5.5.1960). Zhong-Gong Tianjin shiwei Fangbing lanshi bangongshi: quanyu jieshu fangbing lanshi bangongshi gongzuo baogao.
TCA 401206800-X0085-Y-000055-016 (30.5.1960). Hu Yunchang: Jixu kaizhan zengchan jieyue yundong geng haode wancheng shichang gongying renwu.
TCA X0096-C-000026-012 (15.6.1959). Song Jingyi fushizhang zai Tianjin Shi zhaokaide qiucai chan-xiao huiyishang suo zuode dahui baoao.
TCA X0095-Y-000753-038 (28.6.1960). Tianjin Shi Liangshiju: guanyu 7 yuefen gongying anpaide tongzhi 487 hao.
TCA 401206800-X0085-Y-000055-016 (September 1960). Hu Yunchang zai shiwei dahui shangde jianggao: guanyu Tianjinqu yasuo liangshi xiaoliang anpai renmin shenghuo deng jixiang gongzuode yijian.
TCA X0003-Y-008983/28 (5.10.1960). Zhong-Gong Tianjin shiwei: shiwei pizhuan shiwei quanzu nongcun laodong mengliu bangongshi "Guanyu jianjue zhizhi nongmin mengmu wailiude yijian".
TCA X0003-Y-008983 (16.10.1960). Zhong-Gong Tianjin shiwei: guanyu tuoshan anpai zhiyuan nongye huanxiang renyuande tongzhi.
TCA X0095-C-000793-016 (28.11.1960). Tianjin Shi Fushipinju tonggao.
TCA X0095-C-000793-016 (10.12.1960). Tianjin Shi fushipinju: dui fuzhongbing gongying qingkuang jianyao baogao.
TCA X0003-Y-008977 (19.12.1960). Zhong-Gong Tianjin shiwei 420 hao: shiwe pizhuan shi weiju dangzu guanyu Cang Xian renkou simang qingkuang baogao.
TCA X0095-Y-000753-038 (20.12.1960). Tianjin Shi Liangshiju: guanyu 1961 nian 1 yuefen liangshi gongying pinzhong anpaide tongzhi 879 hao.
TCA X0196-C-001086-004/8-15 (collection of documents).
TCA X0196-C-001086-037 (22.12.1960). Tianjin Shi renmin weiyuanhui waishi bangongshi guanyu Nankai daxue jiangshi Liu Di yingzhu jiao Su Xihe yaoqiu zhaogu wenti 472 hao.
TCA X0095-C-000793-016 (24.12.1960). Zhong-Gong Tianjin shiwei Shenghuo bangongshi guanyu jiaqiang fuzhong huanzhe yingyang buzhu gongying gongzuode tongzhi 3 hao.
TCA X0095-C-000793-016/78 (24.12.1960). Zhong-Gong Tianjin shiweihuiyuan Shenghuo bangongshi guanyu jiaqiang fuzhong huanzhe yingyang buzhu gongzuode tongzhi.
TCA X0196-C-001086-004/18 (10.1.1961). Shizheng er gongsi gongzuoqu: Shizheng er gongsi gongzuoqu guanyu shiyong xiaoqiuzao zhiliao fuzhongde chubu shiyan xiaoguo.
TCA X0196-C-001086-037 (18.1.1961). Zhong-Gong Tianjin shiwei Jijianbu shenghuo bangongshi: guanyu fuzhong bingren shouti xiuyang shixing banfa.
TCA X0196-C-001086-004/38 (22.1.1961). Shiwei Jijianbu Shenghuo bangongshi guanyu zengjia dui fuzhongrende guanli, renzhen shixing jizhong liaoyangde qingkuang jianbao.
TCA X0003-C-010770/29 (4.2.1961). Hebei Sheng Shenghuo bangongshi guanyu Hebei, Tianjin, Nankai san ge daxue fangzhi fuzhong bing jingyande baogao.

TCA X003-C-010877/28 (4.2.1961). Heibei Shenghuo bangongshi dianbao.
TCA X0196-C-001086-004/35 (2.12.1961). Tianjin Shi weiyuanhui Shenghuo bangongshi guanyu bing ruo zengjia yingyang gongyingde tongzhi.
TCA X003-C-010877 (13.2.1961). Heibei Shenghuo bangongshi guanyu Wei Xian Fuzhong bingren houguo zao can laodong zao simangde tongbao.
TCA X0198-Y-000984-002 (28.2.1961). Guanyu zhong-xiaoxue he you'ereryuan hanjia qijian shi-sheng fuzhong bing fangzhi qingkuang baogao.
TCA X0196-C-001086-037 (27.3.1961). Shiwei Daishipin bangongshi: Guanyu shinei daishipin shengchan qingkuang ji 3-6 yuefen anpai yijian.
TCA X0196-C-001086-004/21 (14.4.1961). Zhong-Gong Tianjin Shi weihuiyuan Shenghuo bangongshi guanyu fangzhi fuzhong gongzuo zongjie he jinhou yijian.
TCA X0095-C-000793-016 (28.4.1961). Tianjin Shi Fushipinju guanyu tiaozheng ying'er, ertong, yun, chan, ru fu deng fushipinde tongzhi 49 hao.
TCA X028-C-001101-023/94 (16.5.1961). Hebei Sheng Shangyeting / Weishengting guanyu baozheng gongying fuzhong bingren yingyang zhiliao yong tangde lianhe tongzhi.
TCA X0196-C-001086-004/35 (29.11.1961). Tianjin Shi weiyuanhui Shenghuo bangongshi chengren ji ertong bing-ruo qingkuang yuebao biaode tongzhi.
TCA 401206800-X0283-C-001370-015 (28.12.1963). Tianjin Shi shengchan jiuzai weiyuanhui: Hebei sheng Weishengting guanyu zaiqu yiyao jiujikuanzhong jige wentide yijiande baogao.
TCA 401206800-X0283-C-001370-015 (11.1.1964). Tianjin Shi shengchan jiuzai weiyuanhui: Dui zaiqu chunjie gongying wentide tongzhi.
TCA 401206800-X0283-C-001370-015 (15.1.1964). Tianjin Shi shengchan jiuzai weiyuanhui: Guanyu zhaokai fenhongqu he anzhiqu she, dui ganbu pin-xiazhongnong daibiaohuiyide tongzhi.
TCA 401206800-X0283-C-001370-015 (29.1.1964). Zhong-Gong Hebei shengwei, Hebei renminweiyuanhui: guanyu zaiqu chunjie qi shengchan jiuzai gongzuo anpaide tongzhi.

Index

For Product Safety Concerns and Information please contact our EU
representative GPSR@taylorandfrancis.com
Taylor & Francis Verlag GmbH, Kaufingerstraße 24, 80331 München, Germany

www.ingramcontent.com/pod-product-compliance
Ingram Content Group UK Ltd.
Pitfield, Milton Keynes, MK11 3LW, UK
UKHW021721240926
475702UK00029BA/1736
9781138476172